SpringerBriefs in Applied Sciences and Technology

More information about this series at http://www.springer.com/series/8884

Bahram Barati · Iraj Sadegh Amiri

In Silico Engineering of Disulphide Bonds to Produce Stable Cellulase

 Springer

Bahram Barati
University Technology Malaysia
Skudai
Malaysia

Iraj Sadegh Amiri
University of Malaya
Kuala Lumpur
Malaysia

and

Laser Center, Ibnu Sina ISIR
Universiti Teknologi Malaysia (UTM)
Skudai, Johor Bahru
Malaysia

ISSN 2191-530X   ISSN 2191-5318 (electronic)
SpringerBriefs in Applied Sciences and Technology
ISBN 978-981-287-431-3   ISBN 978-981-287-432-0 (eBook)
DOI 10.1007/978-981-287-432-0

Library of Congress Control Number: 2015931419

Springer Singapore Heidelberg New York Dordrecht London

© The Author(s) 2015
This work is subject to copyright. All rights are reserved by the Publisher, whether the whole or part of the material is concerned, specifically the rights of translation, reprinting, reuse of illustrations, recitation, broadcasting, reproduction on microfilms or in any other physical way, and transmission or information storage and retrieval, electronic adaptation, computer software, or by similar or dissimilar methodology now known or hereafter developed.
The use of general descriptive names, registered names, trademarks, service marks, etc. in this publication does not imply, even in the absence of a specific statement, that such names are exempt from the relevant protective laws and regulations and therefore free for general use.
The publisher, the authors and the editors are safe to assume that the advice and information in this book are believed to be true and accurate at the date of publication. Neither the publisher nor the authors or the editors give a warranty, express or implied, with respect to the material contained herein or for any errors or omissions that may have been made.

Printed on acid-free paper

Springer Science+Business Media Singapore Pte Ltd. is part of Springer Science+Business Media
(www.springer.com)

Preface

Cellulase is an industrial enzyme which has significant applications in biofuel production and cellulosic waste management. Cellulase 7a from *Trichoderma reesei* is the most efficient enzyme in biohydrolysis of cellulose. In order to improve its thermostability it can be engineered by many approaches, such as hydrophobic interactions, aromatic interactions, hydrogen bonds, ion pairs, and disulfide bridge creation. In this study, introduction of disulfide bonds into the enzyme was chosen as an approach to achieve this aim. According to disulfide by design software, potential residues for creating disulfide bonds were identified. Accordingly, nine residues are mutated to cysteine, and as a result, five disulfide bridges were created. In order to study conformational stability of all the mutated proteins, molecular dynamic simulation was run for 20,000 ps by using GROMACS software. Root-mean-square deviation (RMSD), Root-mean-square fluctuation (RMSF), and Radius of gyration calculations were used for analysis the results. From the result it can be concluded that disulfide bridges can improve the stability of the protein if they are introduced in specific sites. In addition, more disulfide bridges make the protein more stable by increasing the compactness.

Contents

Chapter 1
Introduction of Cellulase and Its Application

Abstract Energy fuel consumption has been raised considerably in past decades and encouraged researchers to seek new and sustainable source for produce fuel. Cellulosic material are the most abundant natural resource in the earth and can be used for biofuel production to be specific bio ethanol. Ethanol production processes consist of four general steps including cellulose hydrolysis, fermentation, separation of biomass impurities, and purification of ethanol. Cellulose hyrolysis by enzymatic approach cause considerable attention to cellulase as the only enzyme in this process to be engineered. In fact this procedure employs three main group of cellulase including: endoglucanases, exoglucanases and β-glucosidases. In addition these enzyme can be obtained from fungi and bacteria. As rate of chemical reaction increases by rising the temperature, producing stable enzyme in higher temperature (thermostable enzyme) can increase the efficiency of the conversion.

Keywords Fuel · Cellulosic material · Cellulase enzymes · Cellulosic hydrolysis · Themostable enzyme

1.1 Biofuel as a Suitable Alternative for Fossil Fuel

Total energy consumption has been increasing rapidly due to both growth of human population and industries. As a result of this huge energy demand the fuel price has been increasing from 20 to 130 dollars over last 20 years. It is a well-known fact that the combustion of fossil fuel causes environmental pollution by releasing sulphur, nitrogen oxides, carbon monoxide, and suspended particles such as smoke oil and fly ash. Biofuel, on the other hand, can be environmentally friendly depending on its type and source, such as biodiesel from waste is more environmentally friendly than that of soy. Bioethanol produced by fermentation of sugars can be considered as substitute environmentally friendly fuel [1].

Ethanol production processes consist of four general steps including cellulose hydrolysis, fermentation, separation of biomass impurities, and purification of

© The Author(s) 2015
B. Barati and I. Sadegh Amiri, *In Silico Engineering of Disulphide Bonds to Produce Stable Cellulase*, SpringerBriefs in Applied Sciences and Technology, DOI 10.1007/978-981-287-432-0_1

ethanol. Cellulose hyrolysis is carried out by either chemically or enzymatically. In enzymatic process, cellulose is hydrolysed by cellulase [2]. Compared with conventional acid or alkaline (chemical) hydrolysis, enzymatic hydrolysis requires less utility cost as it is normally carried out at mild conditions [3]. As a result, over the last 10 years use of cellulose in producing biofuel has been of great interest [4].

The operative modification of cellulase to sugar involves three types of enzymes (1) endoglucanases (EC 3.2.1.4), glucose and cello-oligosaccharides which have the same structure with cellulose but are shorter in length and are resulted from randomly cut cellulose chains by its activity [5], (2) exoglucanases (EC 3.2.1.91), in order to yield cellobiose (cellobiose is consists of two glucose molecules linked by a $\beta(1 \rightarrow 4)$ bond [6]) which attack exotically to the reducing or non-reducing end of celluloses, and (3) β-glucosidases (EC 3.2.1.21), in order to form glucose hydrolyses cellobiose and cello-oligosaccharides.

Varieties of microorganisms can make cellulolytic enzymes such as bacteria and fungi which can be anaerobic, aerobes, mesophiles and thermophiles. *Trichoderma* species was known as one of the important commercially cellulase producer [7]. Cellulases from the fungal genus *Trichoderma* have attracted significant attention because of their potential to efficiently hydrolyse crystalline cellulose. *Trichoderma* cellulase content is efficient due to enzymatic mixture: two or more exoglucanases or cellobiohydrolase Cel7A (cellobiohydrolases1 or CBH1), and five endoglucanases (EG1–EG5) Cel6A (cellobiohydrolases2 or CBH2), several β-glucosidases, and hemicelluloses [8]. *Trichoderma reesei* also has the ability to produce comparatively large amount cellulase enzyme. It is reported that more than 100 g of cellulases per litter of culture broth is produced by the fungus, whereas a bacteria produce significantly less amount, only few grams per litter [9].

Cellobiohydrolase I (Cel7A) is known as one of the most important exoglucanases and is largely produced by *Trichoderma reesei*. Cel7A contains several domains, big catalytic domain which containing a minor binding unit joined to each other by a linker peptide plus an active site tunnel (Fig. 1.1). The

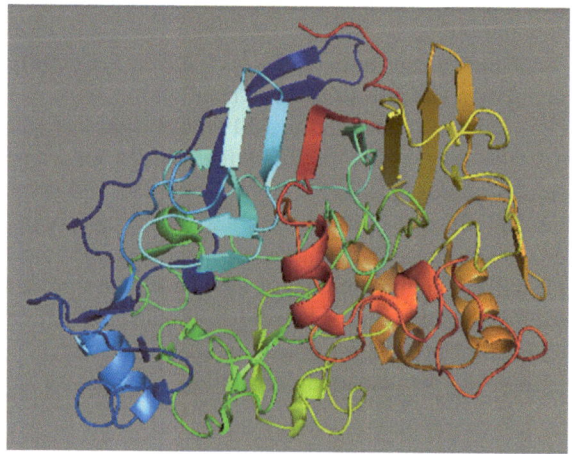

Fig. 1.1 The structure of cellulase 7a from *Trichoderma reesei* obtained from protein data bank (PDB) by accession number of 1DY4

carbohydrate binding domain both identifies and binds to the cellulose surface. Through unknown mechanism, for hydrolysis to cellobiose a single cello dextrin chain is detached from the crystal and nourished into the active site tunnel of the catalytic unit [10].

1.2 Enzyme Stability Is One of the Limits in Biofuel Production

Cellulosic biomass is a cheap feedstock for bioethanol conversation there are some technical challenges still yet to be solved, such as low enzymatic hydrolysis efficiency in cellulase production, reducing numbers of steps in the process, recycling the input material [11]. Therefore, one of the important solutions to these challenges relays on the development of greatly efficient and profitable biocatalysts in fermentation process. For thermo stable enzymes in the hydrolysis of cellulosic materials, there is a number of possible benefits such as greater activity and stability, which leads to enhancement of protein half-life and its recyclability. The first two characteristics would probably increase the activity of the enzymatic hydrolysis even at the range of 45–50 °C [12]. Therefore, thermo-stable enzymes, if used, would eventually improve better activity. Moreover requirement of enzyme amount decreases, hydrolysis time reduces and, accordingly leads to potentially reduce the cost of hydrolysis. Due to lesser viscosity at higher temperatures thermo stable enzymes would expectedly also allow hydrolysis at greater stability and thus permit further elasticity in the procedure configurations [13].

1.3 A Procedure to in Silico Engineering of Cellulase

The overall aim of the current study is to identify the most stable conformation of Cellobiohydrolase I (Cel7A) with the following objectives:

(1) To identify potential mutation sites that can contribute in stabilizing the protein.
(2) To run molecular dynamics simulation in order to screen stable conformation after mutation.

1.4 Cellulase 7a Selected as a Target

The current study is exclusively computational in nature and a standard molecular dynamics simulation methodology was followed to perform the simulation. All the simulations were run in the departmental computational facilities. The extent of conformational stability was measured in terms of root means square deviation (RMSD), radius of gyration (Rg), and root means square fluctuations (RMSF). Based on these measurements and 3D of the conformations, the results are assessed and discussed, Catalytic core of Cellulase 7a from *Trichoderma reesei* containing 434 residues was chosen as scope of study. Accordingly structure was obtained from Protein Data Bank or PDB from the following website http://www.rcsb.org/pdb/home/home.do with accession number of 1DY4. The structure of the catalytic domain can be seen in the Fig. 1.1.

1.5 Advantages of Cellulose in Silico Engineering

Cellulase is known as industrial enzyme, it has important application in lignocellulosic conversion which is used in biofuel production and cellulosic waste management. Although this enzyme is, to some extend thermo stable (optimum temperature 50 °C) this feature can be improved by protein engineering. Molecular dynamic simulation as computational method is used by using rational design approach since it is economical and quicker.

References

1. S.G. Rashmi Kataria, Saccharification of Kans grass using enzyme mixture from Trichoderma reesei for bioethanol production. Bioresour. Technol. **102**, 9970–9975 (2011)
2. L.R. Lynd, Overview and evaluation of fuel ethanol from cellulosic biomass: technology, economics, the environment, and policy. Ann. Rev. Energy Env. **21**, 403–465 (1996)
3. J. Xu, Z. Wang, J.J. Cheng, Bioresource technology Bermuda grass as feedstock for biofuel production: a review. Bioresour. Technol. **102**(17), 7613–7620 (2011)
4. G. Bayram Akcapinar, O. Gul, U.O. Sezerman, From in silico to in vitro: modelling and production of Trichoderma reesei endoglucanase 1 and its mutant in Pichia pastoris. J. Biotechnol. **159**(1–2), 61–68 (2012)
5. M. Umemura, Y. Yuguchi, T. Hirotsu, Interaction between cellooligosaccharides in aqueous solution from molecular dynamics simulation: comparison of cellotetraose, cellopentaose, and cellohexaose. J. Phys. Chem. A **108**(34), 7063–7070 (2004)
6. S.G. Lee, J.I. Choi, W. Koh, S.S. Jang, Adsorption of β-d-glucose and cellobiose on kaolinite surfaces: density functional theory (DFT) approach. Appl. Clay Sci **71**, 73–81 (2013)
7. J.R. Cherry, A.L. Fidantsef, Directed evolution of industrial enzymes: an update. Curr. Opin. Biotechnol. **14**(4), 438–443 (2003)
8. I. Ng, S. Tsai, Y. Ju, S. Yu, T.D. Ho, Dynamic synergistic effect on Trichoderma reesei cellulases by novel b-glucosidases from Taiwanese fungi. Bioresour. Technol. **102**, 6073–6081 (2011)
9. A. Amore, V. Faraco, Potential of fungi as category I consolidated bioprocessing organisms for cellulosic ethanol production. Renew. Sustain. Energy Rev. **16**(5), 3286–3301 (2012)
10. X. Zhao, T.T.R. Rignall, C. McCabe, W.S. Adney, M.E. Himmel, Molecular simulation evidence for processive motion of Trichoderma reesei Cel7A during cellulose depolymerization. Chem. Phys. Lett. **460**(1–3), 284–288 (2008)

11. M.M. Galbe, A review of the production of ethanol from softwood. Appl. Microbiol. Biotechnol. **59**, 618–628 (2002)
12. L. Viikari, A. Terhi, Thermostable enzymes in lignocellulose hydrolysis. Adv. Biochem. Eng. Biotechnol. **108**, 121–145 (2007)
13. L. Viikari, J. Vehmaanperä, A. Koivula, Lignocellulosic ethanol: from science to industry. Biomass Bioenergy **46**, 1–12 (2012)

Chapter 2
Literature Review of Cellulase and Approaches to Increase Its Stability

Abstract Cellulose is now as an important composite of lignolosic material. Cellulose is made up of D-glucose molecules linked together by β-1, 4-glucosidic bonds have complex and diverse structures. Although cellulase has several applications in brewery, animal food, and laundry, pulp and paper industries, its application in bioethanol production attracted many researchers recently. Bioconversion of cellulose to glucose is done by different cellulase enzymes extracted from verity organisms. *Trichoderma reesei* is a well-known fungi species which produce very efficient cockatiel of cellulases. One of the key enzymes in this procedure is cellobiohydrolase which attackes to the end sides of cellulose. However, due to high cost of enzymes bio ethanol is not commercialized yet. One approach to overcome this obstacle is to lower enzyme usage by increasing its stability and efficiency, the most common way to enhance enzyme stability is introducing disulfide bonds. Rational protein engineering tools helped to design more stable enzyme to decrease cellulase production. Recently by advances in computer science the protein can be computationally engineered and the effect can be simulated prior to any lab experiment.

Keywords *Trichoderma reesei* · Cellobiohydrolase · Cellulose structure · Bioconversion · Protein engineering

2.1 Cellulose and Cellulases Structure

Cellulose despite a typical chemical arrangement made up of D-glucose molecules linked together by β-1, 4-glucosidic bonds have complex and diverse structures. The polymeric chain which is linear may have more than 10 thousand insoluble glucose molecules. The chains are not isolated, but, instead, stranded by to each other in a parallel style to form crystalline micro fibrils. Micro fibrils differ on the origin, the size and the crystallinity. Moreover, physical treatments can disturb crystallinity and the grade of polymer production as well [1].

© The Author(s) 2015
B. Barati and I. Sadegh Amiri, *In Silico Engineering of Disulphide Bonds to Produce Stable Cellulase*, SpringerBriefs in Applied Sciences and Technology, DOI 10.1007/978-981-287-432-0_2

In the natural world and in the area of industry cellulases are highly essential enzymes, since they have a significant activity in the carbon cycle globally which are able to break down cellulose that is not soluble to sugars molecules that are soluble. The most varied class of enzymes that catalysing the hydrolysis of a substrate are cellulases. Cellulases can be classified into three main classes including: endoglucanases, exocellulases, and processive endoglucanases which are dissimilar structurally and functionally. The cellulases have a catalytic domain (CD) and a substrate-binding unit which this property is similar to other enzymes hydrolysing substrates that are not soluble. Three diverse methods are exist used by microorganisms that degrade cellulose: secretion of a set of free cellulases that work together (synergistic enzyme), production of cellulosomes which are several enzyme complexes, and an unidentified mechanism that does not need processive cellulases that were critical in the two mentioned way [2].

2.2 Applications of Cellulase

Broad elementary and practical investigation for the period of the 1970s and 1980s proved that the enzyme-induced bio-conversion of lignocellulose to sugars that can be solved was relatively challenging. Nevertheless, studies on hemicellulases, cellulases and pectinases shown their biotechnological applications in different industries, such as brewery, food, animal feed, textile and laundry, pulp and paper, agriculture and wine [3].

2.2.1 Application of Cellulase in Biofuel Production

With the unavoidable reduction of the world's petroleum source, a growing universal attention to substitute, resources of energy that is not based on petroleum has been increasing. Over the past two decades years as petroleum provides 97 % of the energy spent for transportation governments and industries global has been dynamically recognizing, rising and commercializing technology for substitute fuels for transportation. There is an increasing use of ethanol as fuel for transportation in the United States. The ethanol produced during fermentation can be expensive and government subsidy is required to keep the price down. Almost all ethanol fuel is created throughout fermentation of sucrose in the Brazil or corn glucose in United States, but any other country with a noteworthy budget based on agriculture are able to use present technology for making petroleum ethanol. This is likely as, throughout the past 20 years, technology for ethanol making from non-food-plant resources has been established to the purpose that massive amount of fuel will be a fact in the few years later. Therefore, agricultural products including wheat or rice straw, corn Stover (corn cobs and stalks), waste sugar cane, the paper potion of municipal waste, forestry and paper mill castoffs, and dedicated energy crops together are called 'biomass' can be transformed to ethanol that can be used as a fuel [4].

The main problem for commercialization of ethanol made by fermentation is much costly compared to the native gasoline. Latest rises in the comprehensive value of oil seem to be aiding to bridge the fee breach among gasoline and ethanol. The price of ethanol is related, in part, to the unpreventable loss of partial of the carbon throughout microorganism's fermentation of sugars. Though ethanol production from cane sugar is a relatively easy procedure, trouble raises when these procedures need enzymes to hydrolyse starch to glucose former to fermentation when ethanol is made from corn or wheat starch. Ethanol production from biomass involves even broader treating to release the polymeric sugars in hemicellulose and cellulose that account for 20–35 % and 23–53 % of biomass, individually. Cellulose is a polymer made by beta-linked glucoses, while hemicellulose is a meaningfully diverged chain of arabinose and xylose that has glucose, mannose and galactose as well. Usually acids hydrolysis of these carbohydrate polymers such as cellulose, amylose, hemicellulose and starch (contributed either by the biomass or added externally) and enzymes. Concentrated sulphuric acid used in the hydrolysis of the biomass carbohydrates. The sugars are parted from the acid once hydrolysed, and then are fermented to ethanol. The pre-treatment techniques aim to diverse the lignin matrix and carbohydrates while chemical demolition of sugars throughout fermentation is vital for producing ethanol. Improvement of a faultless pre-treatment technique is problematic, given that 'biomass' contains sources such as hardwood and softwood trees, agricultural remains products like corn Stover, and paper that cannot be recycled. Scientists have tested numerous pre-treatment procedures such as steam explosion and hot water, and also alkaline and solvent pre-treatments and many valued shapes of acid pre-treatment regarding to the existence of various feedstocks. Newly the yields and kinetics of the countless acid-based batch and flow-through techniques have been analysed, and obviously the flow-through processes make greater sugar yields available and grounds less sugar degradation, but consequences an additional sugar solution. The pre-treatment procedure is considered simply to start the decay of the biomass and partly hydrolyse the polymers of carbohydrates, in order to making them ready to enzymatic attack. For the first time cellulases were used in a sequential procedure (pre-treatment → cellulase hydrolysis → ethanol fermentation). On the other hand, the saccharification (procedure of depredating a complex carbohydrate such as starch or cellulose into its subunits) and fermentation (SSF) procedure affords important reduction in expenses because cellulase hydrolysis happens throughout glucose fermentation. The procedures presently used include fermentation of all biomass sugars in a simultaneous saccharification cofermentation (SSCF) procedure [4]. In order to prepare substrate for fermentation pre-treatment technique is required which might use several ways for treating biomass. Recently acid based pre-treatment are use although it was very challenging when biomass used as substrate of fermentation.

2.3 *Trichoderma reesei* Known as the Crowned King of Cellulolytic Fungi

The finding of the fungus *Trichoderma reesei* during the period of World War II, when the allied militaries in the South Pacific hurt the quick damage of cotton exhaustions and camps. The unseen 'enemy' was recognized as a tiny fungus *T. viride* strain QM 6a. Broad investigations of this fungus were conducted, causing the choice of hyper cellulolytic fungal mutants, with the high-yielding strains *T. viride* QM9414 and MCG77. In 1977, the fungus was give new name *T. reesei* in honour to E.*T. Reese*, who discovered it in the 1940s. The report of Mandels and Sternberg (in 1976) express information on more than 14,000 discovered and screened fungi by seeing their cellulase activity; no serious competitors to *T. reesei* were found. The 'king' of cellulolytic fungi was at the end of the day crowned. Diverse cellulase hyperproducing strains of *T. reesei* have been settled; the *RUT C30* strain is one of the most powerful and best characterized strains, and it has become a reference strain among *T. reesei* high cellulase producers [5]. A surprisingly pitiable repertoire of genes for cellulases and hemicellulases has discovered by the genome sequencing of *T. reesei*. Though ten cellulases including eight endoglucanases and two cellobiohydrolases belonging to different glycoside hydrolase families have been identified in the *T. reesei* genome, only four main cellulases are generally secreted in remarkable amounts by this fungus: CBH I (Cel7A), CBH II (Cel6A), EG I (Cel7B) and EG II (Cel5A) [6]. These normally characterize up to 90–95 % of the whole secreted protein, CBH I making up 50–60 % and CBH II 20 % of the whole cellulases.

A number of hyper thermophilic and thermophilic bacteria live at great temperatures that can go above 100 °C. These microbes are a latent basis of extremely thermo stable cellulases and other enzymes, which are improved, appropriate to strict progression environments and might take greater exact activities. Although bacterial cellulases have interesting and beneficial features, bacteria cannot be compared to mutant strains of fungi by the level of protein production. This feature might come to be the major difficulty during using cellulases from bacteria in the assembly of second-generation biofuels. As a matter of fact, cellulases from bacteria with useful properties can be heterologously expressed in fungi as members of fungal simple enzyme systems. For consolidated bioprocessing, specific bacterial species are suitable precisely the straight microbial transformation of lignocellulosic biomass to ethanol, butanol, organic acids and further valuable yields (Fig. 2.1).

2.4 Catalytic Mechanisms of Glycoside Hydrolases

By acid-base catalysis glycoside hydrolases cut glycosidic bonds. Catalysis can be accomplished with either inversion or net retention of the anomeric configuration (when two molecule have similar formula but only the conformation at hemiketal

Fig. 2.1 In this microscope
image of the fungus
Trichoderma reesei, proteins
are stained *red*, and *white*
chitin, a component of the cell
walls, is stained *blue* (http://
machinedesign.com/news/
using-fungus-make-fuel)

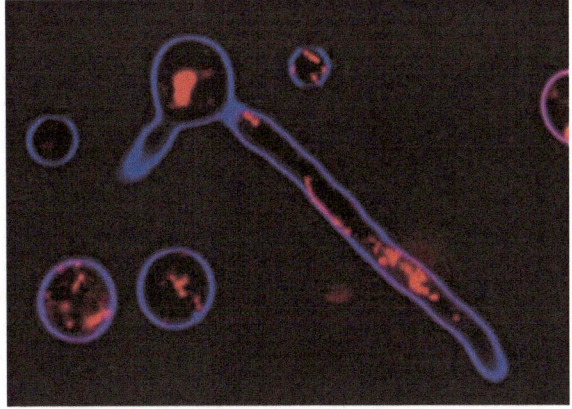

carbon is different) of the substrate (Fig. 2.2). Inversion is a typical shift reaction. A
catalytic acid provides protonic support to a group that leaves (glycoside hydroxyls
have a high pK_a and thus make poor leaving groups) at the same time, in order to

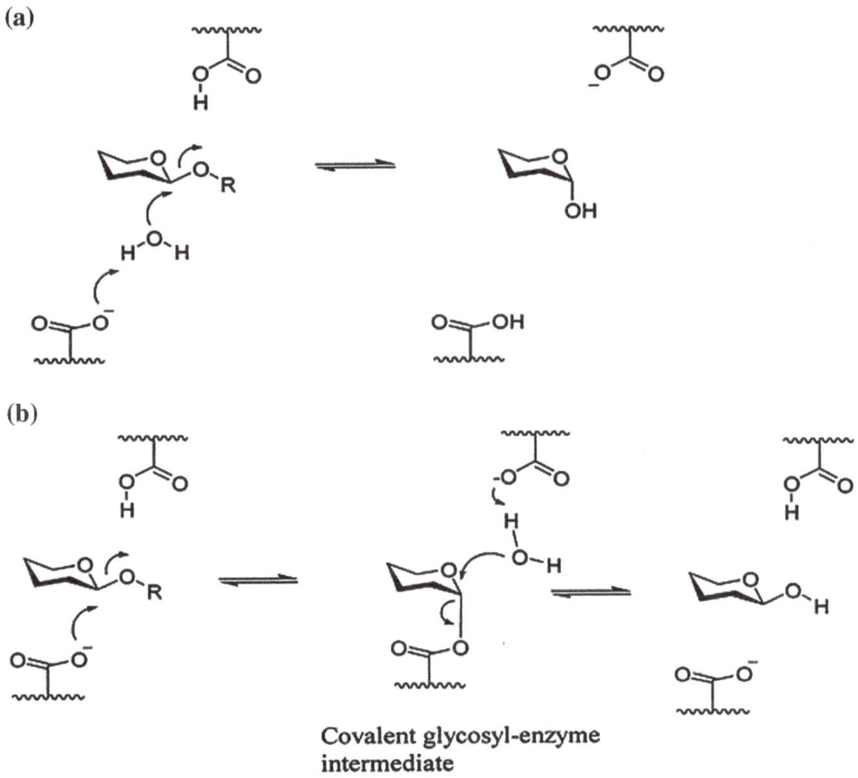

(a)

(b)

**Covalent glycosyl-enzyme
intermediate**

Fig. 2.2 Inversion and retention process. **a** Inversion, **b** retention

discharge proton from a water molecule for nucleophilic replacement at the anomeric core catalytic base. The base and acid both are normally placed some 7–13 Å separately in order to provide somewhere to stay the nucleophilic water 'below' the pyranoside ring. In many cellulase systems, the identification of the catalytic base remains controversial.

The mechanism of retention is a binary movement basically as defined by Koshland in 1953. By oxacarbenium ion-like conversion conditions a covalent glycosyl-enzyme intermediate is made, and then hydrolysed. This involves two vital residues, an enzymatic nucleophile and a catalytic acid/base which first serves as a typical Brønsted acid (the atom that donates hydrogen according to Bronsted-Lowry definition), giving proton to the leaving group to support departure then functions as a base, lose proton from the incoming water nucleophile for the second step. The nucleophile and acid/base are always found some 5–6 Å separately on all systems studied to date. It is worth commenting that, the stereochemistry of catalysis is preserved within each family as catalytic mechanism is dictated by the place of functional groups on the protein. Both inverting and retaining cellulases are identified [7].

2.5 An Introduction to Cellulase Family 7

Cellulases from family 7 (family 7 is a family of glycoside hydrolases which is classified among glycoside hydrolases, based on sequence similarity and carry out catalysis with net retention of configuration), as explained before. From this family three-dimensional (3D) structure of several cellobiohydrolases and endoglucanases are identified. The 3-D structure contains a double β-sandwich. Alike to family 6, the structural difference among the cellobiohydrolases and the endoglucanases is exposed in the size and nature of the loops neighbouring the substrate binding sites: these loops are stretched and form a sealed off tunnel in the cellobiohydrolases.

The 3-D structures of both cellobiohydrolases (CBH I and CBH II) and endo-glucanases from family 7 exposed a 'trio' of carboxylates in the active centre. By mutation of the residues Glu212-Asp214-Glu217, to their isosteric amide counterparts, the functions of them (formerly CBH I), were investigated in *T. reesei* Cel7A. The three point mutations meaningfully reduced the catalytic activity of the enzyme, though all keep a number of residual activities. On the minor chromophoric substrate 2-chloro-4-nitrophenyl β-lactoside the k_{cat} (catalytic constant) standards were fall down to 1/2,000, 1/85 and 1/370 of the native activity, respectively, whereas the K_m (Michaelis constant which shows enzyme's affinity for a substrate) (values stayed basically unaffected. On insoluble bacterial microcrystalline cellulose (BMCC) no noteworthy activity was distinguished for the Glu212Gln and Glu217Gln mutants, whereas the Asp214Asn mutant retained residual activity. From many studies it is now obvious that the two glutamates work correspondingly as the catalytic nucleophile and catalytic acid/base (Table 2.1).

Table 2.1 Catalytic residue equivalences in family 7 [1]

Enzyme	Organism	Catalytic acid/base	Catalytic nucleophile
Cel7A (CBH I)	*T. reesei*	Glu217	Glu212
Cel7B (EG I)	*T. reesei*	Glu201	Glu196
Cel7B (EG I)	*H. insolens*	Glu202	Glu197
Cel7B (EG I)	*F. oxysporium*	Glu202	Glu197

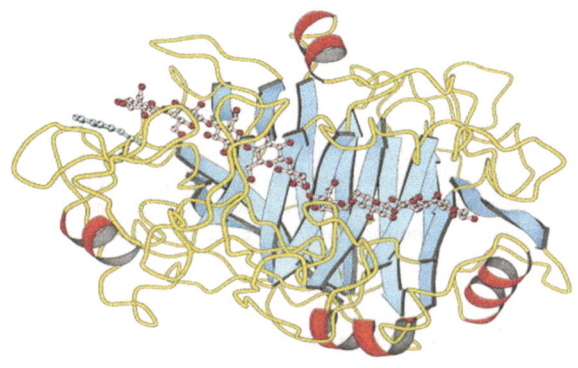

Fig. 2.3 The CBH I catalytic domain with a cello oligomer bound secondary-structure elements are colored, *blue arrows* for β strands; *red spirals* for α helices; *yellow coils* loop areas. The cello oligomer is highlighted in *pink* as a ball-and-stick object [8]

Cellulose chain was bound alongside the 50 Å substrate binding tunnel. Active site variations Glu212Gln and Glu217Gln were crystallized together with cello-oligosaccharides and the structures disclose glucose moieties covering sub sites as well. The binding manner witnessed matches to that predicted throughout productive binding of a cellulose chain and provides the theory that hydrolysis by *T. reesei* Cel7A continues from the reducing end of a cellulose chain (Fig. 2.3).

2.6 Cellulose Biodegradation Procedures

Cellulose biodegradation begins by binding of cellulase to cellulose by binding sites which recognize reducing end of cellulose then degradation of cellulose begins in catalytic tunnel and hydrolysis occurs. Summarized interaction of the cel7a can be observed in the Fig. 2.4.

2.7 Protein Engineering as a Solution

Enzymes are naturally designed protein that catalysts and are able to work in physiological states. On the other hand, biotransformation involves the enzymes in conditions that may proceed considerably from physiological condition. It is very

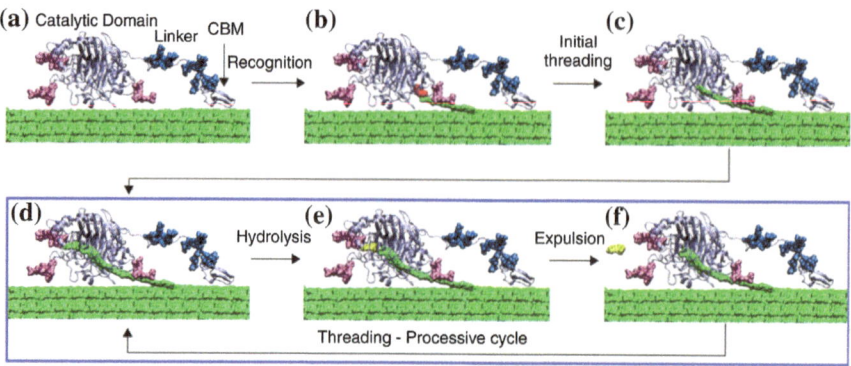

Fig. 2.4 a Cel7A binding to cellulose, **b** recognition of a reducing end of a cellulose chain, **c** beginning threading of the cellulose chain into the catalytic tunnel, **d** threading and formation of a catalytically active complex, **e** hydrolysis in a processive cycle and **f** product expulsion and threading of another cellobiose (shown in *yellow* in **e** and **f**). Image reproduced with publisher's permission [9]

difficult to engineer enzyme that would optimally perform under extreme conditions. Alternatively, many enzymes are capable of catalysing a number of substrates and/or transforming some substrates, they are more promiscuous catalysts [10].

In the decades since the inception of protein engineering three approaches have been taken. Early work was focused on rational design and thereby restricted to enzymes where there was a considerable knowledge of structure. In the next generation, the emphasis was on the development of diversity via random mutagenesis to create large libraries. In the last decade this type of library has developed such that a greater diversity is used as the starting point, also using natural diversity [11].

2.7.1 Rational Design Approach

The latest approach for engineering enzymes is rational protein design (RPD). RPD contains a set of molecular biology methods, for example site-directed mutagenesis (SDM), according to structure of enzyme investigations. The correct reorganization of residues in charge for substrate-enzyme relations, particularly, docking, stability and activity is vital for the use of RPD in a directed way. Occasionally, it is essential to do sequence homology or alanine scanning of linked species mutual with biophysics data. Progresses in the likelihood of the consequence of SDM have been made so, at the same time the effect of mutation on ligand affinity, stability, and pK(a) values can be assessed and predicted for several mutants in one step. Due to the fact that RPD desires the information of the interest enzyme structure and/or its sequence in numerous and linked to species, the crystallography and spectroscopic analysis of numerous enzymes have been a great device to practice in modelling by computer. This method relays on the improvement completed in determination of structure, better modelling rules and new visions into structure–function relationships. Then again, progresses in modelling, mainly calculations of free energy perturbation and molecular dynamics (MD) can calculate the suitable mutations for the development of enzyme selectivity.

On the other hand computer simulation investigations were established for the enhancement of enzyme catalysis. Recently, in MD simulations, estimated free energy and hydrogen bond energy calculations were integrated to reveal the structure–activity interactions. Similarly, the integration of diverse prediction for structure methods can be used in RPD of enzymes as is the case of molecular docking, fragment molecular orbital method (FMO) calculation, Quantitative structure–activity relationship (3D-QSAR), Comparative Molecular Field Analysis (CoMFA) modelling [12].

Since, structural data are not available always, a dissimilar idea as consensus sequence design (CSD) was established. CSD is an interesting idea for making protein more stable, that exploits amino acid conservation in sets of homologous proteins to recognize likely good mutations and it does not rely on the accessibility of data of structure. Data-driven CSD is according to the typical hypothesis that the rate of a certain residue in a multiple sequence alignments (MSA) of homologous proteins contacts with that amino acid's participation to protein stability. On the other hand, its achievement relays on the phylogenetic variety of the sequence set obtainable. The practice of this technique shows that a phylogenetically balanced CSD can lead in mesostable proteins noteworthy stabilization of secondary structural motifs. Enhancing the amino acid alphabet used for making random by taking into account structural details and differences in functions, and considering covariation in the plan procedure are further plans that could be exploited to make the most of stabilization while conserving activity [13].

Newly, RPD by MD simulations was used in thermo stability enhancement without decreasing enzyme performance taking considering properties of protein's surface as a substitute on core of protein features including core packing and cavity filling. Thus, flexible residues in the surface tolerant to mutations are practical objectives for thermo stabilization and that local-interaction stabilization of cavity lining residues by means of the MD technique can be an operative substitute to the typical cavity filling technique. Methods which employ computers can perhaps guide the RPM of proteins making DE more useful. On the other hand, the exploitation of molecular possible tasks can be cooperative to calculate the influences of mutations on protein stability and structure for collections of enzyme variants created in silico. Numerous illustrations of the combination of DE and RD might be found in the literatures to increase stability, adaptation to cold, new activities and creates, substrate specificity using sequence base enzyme remodel between the rests [14].

2.8 Advantages of Enzyme Thermo Stabilization

Enzyme thermo stabilization is very much of interest by industry and research. Keeping the protein stable at temperatures that would normally denature and destabilize it can be a main objective. Or a second demand can be to retain the protein stable at sensible temperatures, but for longer time (increasing the half-life). These two sides of stability required to somehow a different practice to the problem, but typically both effects are found together in thermo stable proteins. Thermo stable enzymes are needed to breaking down numerous organic compounds to working sources of energy. For instance the breakdown of cellulose for the aim of generating economically possible ethanol, a procedure that at present is too costly [15].

2.9 Some Thermo Stabilize Mutations

There are many mechanisms for enzyme thermo stabilization including; engineering: amino acid composition and intrinsic propensity, disulfide bridges, hydrophobic interactions, aromatic interactions, hydrogen bonds, ion pairs, prolines and decreasing the entropy of unfolding, intersubunit interactions and oligomerization, conformational strain release, helix dipole stabilization packing and reduction in solvent-accessible hydrophobic surface, docking of the n and c termini, and anchoring of loose ends, metal binding, nonlocal versus local interactions, post translational modifications and Extrinsic Parameters.

2.9.1 Introducing Disulfide Bonds as a Strategy to Increase Stability

In most extracellular proteins disulfide bonds are present, where they seemingly make stable the native conformation by decreasing the entropy of the unfolded state or by lowering the unfolding level of permanently denatured proteins This procedure property makes disulfide bond cross-linking an smart approach for engineering extra stability for conformation into proteins by site directed mutagenesis [16].

A disulfide bridge that the covalent bond is made have to be broken for a fully unfolded state and the breaking of covalent bonds is an energetically unfavourable procedure. The best of a recent characterized protein containing disulfide bridge has an ideal stability and activity beyond 100 °C, which is a probable gage that the adding of this bond can have potential effect in stabilization in more than 100 °C. Though, due to oxidation, at these temperatures cysteine is quite instable, thus this strategy probably is not used much in nature [17].

2.9.1.1 Mechanism of Disulfide Bond Formation

The disulfide bond in vivo formation mechanisms has been broadly investigated in eukaryotic and prokaryotic organisms in the past 20 years [18]. A sequence of thiol/disulfide replacement responses among cysteine thiolate (S⁻) and oxidizing disulfide bond (S–S) by the nucleophilic attack that are done by enzymes inside the cells commonly results disulfide bond formation. Three classes of chemical interactions are available including: isomerization, reduction, and oxidation, which are shown in Fig. 2.5. A disulfide bond is presented to the protein substrate by responding with the oxidase (Fig. 2.5a). In the same way, a formed disulfide bond can be reduced by the reductase in the protein (Fig. 2.5b). The reorganization of disulfide bond inside a protein molecule can happen through two unlike mechanisms: a two-step procedure with a reduction reaction first (Fig. 2.5b) after an oxidation reaction (Fig. 2.5a) or a single-step isomerization procedure (Fig. 2.5c). Two oxidoreductase catalyzed reactions involved in the two-step mechanism: a reduction of a disulfide bond to two thiols and a catalyzed oxidation of thiols to make fresh disulfide bonds. The single-step mechanism contains only one isomerize which make the mixed

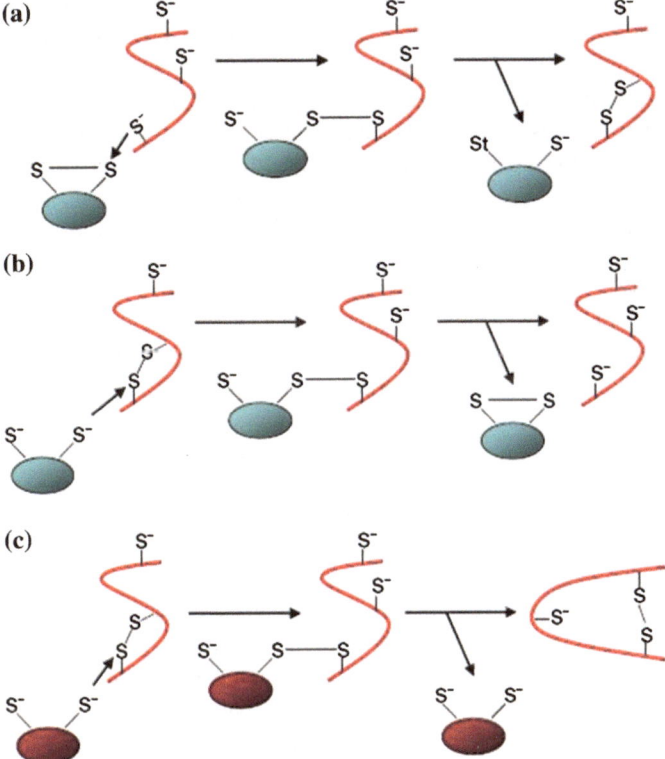

Fig. 2.5 a Oxidation, **b** Reduction, and **c** Isomerization. The oxidoreductase and isomerase enzymes are represented by blue and red ovals, respectively

disulfide bond by isomerizing a disulfide bond. Reliant on the conformation of the protein molecule, the diverse disulfide bond might reply with an close cysteine residue, causing the creation of a novel disulfide bond [19].

2.10 Molecular Dynamic Simulation as a Strong Approach to Evaluating Structure Stability

Since the first molecular dynamics (MD) simulations of a protein described, it has been 25 years. Throughout this period, in the investigation of biomolecules, matching to experimental methods dynamics simulations have become a well-known tool.

Three major areas of application can be found in bio molecular dynamics simulations nowadays. Firstly, MD simulation is used to bring biomolecular structures alive, giving insights into the natural dynamics on different timescales of biomolecules in solution. Secondly, MD simulation affords thermal averages of molecular properties. Based on the ergodic hypothesis, individuals are able to simulate a molecule or group of molecules under a specific environment for a period of time and the macroscopic properties can be calculated by time-averaging all the possible state of the molecules in a system and the set of all the possible states of the molecular system is known as ensemble. This is applied to calculate, for illustration, the specific properties of solutions and the free energy changes for biochemical procedures including ligand binding. Thirdly, MD can discover which conformations of a molecule or a complex are accessible thermally. This method is used for discovering conformational space, for example, in ligand-docking uses. Furthermore, available conformations of a molecule or a complicated are thermally can be find by MD simulation [20]. This way is used for discovering conformational space, for example, in ligand-docking applications. In addition data from experiments used for dynamics calculations, MD can suitably combine these data from experiment with the data about the overall properties of molecular structure that is embodied in the hundreds of factors of a molecular mechanics force field [21].

The MD field catches benefit from the apparently constant developments in computer hardware; simulations that were difficult for past processors can be done these days by normal office workstations.

Over the last decades molecular dynamics simulations have developed from a technique to investigate the dynamics of fluids of solid spheres and Lennard–Jones units to a multipurpose technique to investigate a comprehensive collection of systems at atomic level. The simple technique contains explaining the contacts of all atoms in a certain arrangement by a fairly empirical probable purpose, in this system the energy on whole atoms is able to be calculated and combined in time, by time steps of the order of 1,015 s. A trajectory is resulted as a main outcome and holds the movements of whole atoms in time, more than billions of time steps. Investigation of these movements provides vision into the system that is being considered. Such simulations give entire information of the motions molecules in the system, and hence those and numerical mechanics, contact to thermodynamic properties. There are, however, major restrictions. A significant attention for the simulation application and to evaluation of the accuracy of simulations is the length and time scale on which the procedures of interest happen. Methodical limits related to the precision of the empirical possible work are existing as well, the comparatively shortens of the systems that can be simulated, and difficulties with exactly incorporating noteworthy variables including pH, transmembrane likely alterations, and little ion concentrations. The initial configuration of a simulation might bias the outcomes in unwanted approaches as well [22].

Simulation has many application for example in physics many system can be build up and simulated, the vast amount of date are produced in this field of science. The example of the simulation can be seen in the following papers which study behavior of ring resonators. The ring resonators are suitable for many applications in micro and nano optical communication. Optical soliton is a self-reinforcing solitary wave that maintains its shape while it travels at constant speed. Optical solitons are seen by a cancellation of nonlinear and dispersive effects in the medium which can be a fiber optic. In a Kerr effect medium such as fiber optics, high intensity of light causes a phase delay having similar temporal shape as the intensity. This nonlinear phenomenon occurs for a beam called self-phase modulation (SPM), which is generated by its intensity. Optical Chaos occurs in many nonlinear optical systems. One of the most common examples is a ring resonator. Chaotic behavior has been considered as a nonlinear property in physics, electronics, and communication. When a high-intensity short pulse is coupled to optical fiber, the instantaneous phase of optical pulse rapidly changes through the optical Kerr effect. The SPM and cross-phase modulation (CPM) effects change the phase of the pulse as a function of its intensity. Here, we derive the soliton equations based on solving the nonlinear Schrodinger and Maxwell equations. The main performance characteristics of ring resonators are transmittance, free spectral range, finesse, Q-factor, and group delay, which have been demonstrated [23–33]. It is interesting that these studies have application in biology by rapping bio cells [34].

References

1. M. Schülein, Protein engineering of cellulases. Biochim. Biophys. Acta **1543**(2), 239–252 (2000)
2. A.L. Rodrigues, A. Cavalett, A.O.S. Lima, Enhancement of Escherichia coli cellulolytic activity by coproduction of beta-glucosidase and endoglucanase enzymes. Electron. J. Biotechnol. **13**, 1–9 (2010)
3. M.K. Bhat, Cellulases and related enzymes in biotechnology. Biotechnol. Adv. **18**(5), 355–383 (2000)
4. J.R. Mielenz, Ethanol production from biomass: technology and commercialization status. Curr. Opin. Microbiol. **4**(3), 324–329 (2001)
5. A.V. Gusakov, Alternatives to Trichoderma reesei in biofuel production. Trends Biotechnol. **29**(9), 419–425 (2011)

6. V.V. Morozova, A.V. Gusakov, R.M. Andrianov, A.G. Pravilnikov, D.O. Osipov, A. P. Sinitsyn, Cellulases of Penicillium verruculosum. Biotechnol. J. **5**(8), 871–880 (2010)
7. G.J. Kleywegt, J.Y. Zou, C. Divne, G.J. Davies, I. Sinning, J. Ståhlberg, T. Reinikainen, M. Srisodsuk, T.T. Teeri, T.A. Jones, The crystal structure of the catalytic core domain of endoglucanase I from Trichoderma reesei at 3.6 A resolution, and a comparison with related enzymes. J. Mol. Biol. **272**(3), 383–397 (1997)
8. C. Divne, J. Ståhlberg, T.T. Teeri, T.A. Jones, High-resolution crystal structures reveal how a cellulose chain is bound in the 50 A long tunnel of cellobiohydrolase I from Trichoderma reesei. J. Mol. Biol. **275**(2), 309–325 (1998)
9. G.T. Beckham, Y.J. Bomble, E.A. Bayer, M.E. Himmel, M.F. Crowley, Applications of computational science for understanding enzymatic deconstruction of cellulose. Curr. Opin. Biotechnol. **22**(2), 231–238 (2011)
10. O. Khersonsky, D.S. Tawfik, Enzyme promiscuity: a mechanistic and evolutionary perspective. Annu. Rev. Biochem. **79**, 471–505 (2010)
11. J.M. Woodley, Protein engineering of enzymes for process applications. Curr. Opin. Chem. Biol. **17**(2), 310–316 (2013)
12. Q. Zhang, J. Yang, K. Liang, L. Feng, S. Li, J. Wan, X. Xu, G. Yang, D. Liu, S. Yang, Binding interaction analysis of the active site and its inhibitors for neuraminidase (N1 subtype) of human influenza virus by the integration of molecular docking, FMO calculation and 3D-QSAR CoMFA modeling. J. Chem. Inf. Model. **48**(9), 1802–1812 (2008)
13. J.C. Joo, S.P. Pack, Y.H. Kim, Y.J. Yoo, Thermostabilization of Bacillus circulans xylanase: computational optimization of unstable residues based on thermal fluctuation analysis. J. Biotechnol. **151**(1), 56–65 (2011)
14. A. Illanes, A. Cauerhff, L. Wilson, G.R. Castro, Recent trends in biocatalysis engineering. Bioresour. Technol. **115**(2012), 48–57 (2012)
15. C. Vieille, G. Zeikus, Hyperthermophilic enzymes: sources, uses, and molecular mechanisms for thermostability. Microbiol. Mol. Biol. Rev. **65**(1, 517), 1–43 (2001)
16. O.R. Siadat, A. Lougarre, L. Lamouroux, C. Ladurantie, D. Fournier, The effect of engineered disulfide bonds on the stability of Drosophila melanogaster acetylcholinesterase. BMC Biochem. **7**, 12 (2006)
17. D.B. Volkin, A.M. Klibanov, Thermal destruction processes in proteins involving cystine residues. J. Biol. Chem. **262**(7), 2945–2950 (1987)
18. J. Messens, J.-F. Collet, Pathways of disulfide bond formation in Escherichia coli. Int. J. Biochem. Cell Biol. **38**(7), 1050–1062 (2006)
19. L. Zhang, C.P. Chou, M. Moo-Young, Disulfide bond formation and its impact on the biological activity and stability of recombinant therapeutic proteins produced by Escherichia coli expression system. Biotechnol. Adv. **29**(6), 923–929 (2011)
20. T. Hansson, C. Oostenbrink, W. van Gunsteren, Molecular dynamics simulations. Curr. Opin. Struct. Biol. **12**(2), 190–196 (2002)
21. E. Olkhova, Molecular dynamics simulations and hydrogen-bonded network dynamics of cytochrome c oxidase from Paracoccus denitrificans (2004)
22. W.L. Ash, M.R. Zlomislic, E.O. Oloo, D.P. Tieleman, Computer simulations of membrane proteins. Biochim. Biophys. Acta **1666**(1–2), 158–189 (2004)
23. S.E. Alavi, I.S. Amiri, H. Ahmad, A.S.M. Supa'at, N. Fisal, Generation and Transmission of 3× 3 W-Band MIMO-OFDM-RoF signals using micro-ring resonators. Applied Optics **53**(34), 8049–8054(2014)
24. A. Nikoukar, I.S. Amiri, S.E. Alavi, A. Shahidinejad, T. Anwar, A.S.M. Supa'at, S.M. Idrus, L.Y. Teng, in *Theoretical and Simulation Analysis of The Add/Drop Filter Ring Resonator Based on the Z-transform Method Theory*. Presented at the The 2014 Third ICT International Student Project Conference (ICT-ISPC2014), Thailand (2014)
25. I.S. Amiri, J. Ali, Simulation of the single ring resonator based on the Z-transform method theory. Quantum Matter **3**(6), 519–522 (2014)

26. I.S. Amiri, S.E. Alavi, S.M. Idrus, A. Nikoukar, J. Ali, IEEE 802.15.3c WPAN standard using millimeter optical soliton pulse generated by a panda ring resonator. IEEE Photonics J. **5**(5), 7901912 (2013)
27. I.S. Amiri, S.E. Alavi, H. Ahmad, A.S.M. Supa'at, N. Fisal, Numerical computation of solitonic pulse generation for Terabit/Sec data transmission. Opt. Quant. Electron. (2014)
28. I.S. Amiri, S.E. Alavi, N. Fisal, A.S.M. Supa'at, H. Ahmad, All-optical generation of two IEEE802.11n signals for 2×2 MIMO-RoF via MRR system. IEEE Photonics J. **6**(6) (2014)
29. Optical wired/wireless communication using soliton optical tweezers. Life Sci. **10**(12) (2013)
30. A. Nikoukar, I.S. Amiri, J. Ali, Generation of nanometer optical tweezers used for optical communication networks. Int. J. Innovative Res. Comput. Commun. Eng. **1**(1), 77–85 (2013)
31. S. Alavi, I. Amiri, S. Idrus, A. Supa'at, J. Ali, Chaotic signal generation and trapping using an optical transmission link. Life Sci. J. **10**, 186–192 (2013)
32. I.S. Amiri, F.J. Rahim, A.S. Arif, S. Ghorbani, P. Naraei, Single soliton bandwidth generation and manipulation by microring resonator. Life Sci. J. **10**, 904–910 (2013)
33. I. S. Amiri, A. Nikoukar, A. Shahidinejad, T. Anwar, The proposal of high capacity GHz soliton carrier signals applied for wireless communication. Rev. Theor. Sci. **2**, 2327 (2014)
34. N. Suwanpayak, S. Songmuang, M.A. Jalil, I.S. Amiri, I. Naim, J. Ali, P.P. Yupapin, Tunable and storage potential wells using microring resonator system for bio-cell trapping and delivery. Int. Conf. Enabling Sci. Nanotechnol. (0), 1–2 (2010)

Chapter 3
Methodology of Mutant Creation and Molecular Dynamic Simulation

Abstract In order to introduce disulfide bond to a protein certain pair residues need to be mutated to cysteine. To find the best candidate for mutation computational web server (disulfide by design server) is used then PyMol software is used to mutate proposed residues. Followed that molecular dynamic simulation is run in Gromacs package in following seven steps; mutated PDB files need to be readable for Gromacs therefore a tool named pdb2gmx is used to convert PDB to topology file readable by Gromacs (Topology generation), then an aquatic environment need to be defined (box and solvent define), in order to neutralize the system ions need to be added to system (adding ion), in order to brings atoms in the correct position energy minimization need to be perform (energy minimization), then the temperature and pressure of the system need to be equilibrated (equilibration), after these steps system is ready to be simulated for defined time (MD production) and finally analyzing the MD product trajectory files by using Root-Mean square division (RMSD), Root-Mean square fluctuation (RMSF) radius of gyration (RG) calculations (analysis).

Keywords PyMol · GROMACS · Disulfide by design · Simulation steps · Trajectory file analysis

3.1 Obtaining Protein Structure File

Based on name of the protein, the sequence is extracted from Uniprot database (http://www.uniprot.org/) in order to find the 3D structure of the molecule. As a result the Uniport accession number (G0RVK1) is obtained then BLASTP tool is used to search for structural file (pdb file). Accordingly the pdb file of catalytic core of the interested enzyme is extracted by accession number of 2V3I from Protein Data Bank website (http://www.rcsb.org/pdb/home/home.do) by the resolution of 1.05 Å.

© The Author(s) 2015

B. Barati and I. Sadegh Amiri, *In Silico Engineering of Disulphide Bonds to Produce Stable Cellulase*, SpringerBriefs in Applied Sciences and Technology, DOI 10.1007/978-981-287-432-0_3

3.2 Fixing pdb File

3.2.1 Removing Water Molecule and Ligands

The downloaded pdb file contains coordination of surrounding water molecule and attached ligands, in order to making pure protein structure file all the attached molecules need to be removed.

3.3 Generation of Mutated pdb Files

As of the objectives of this study is to make thermo stable protein. Disulfide bonds were introduced to achieve this aim. In order to make disulphide bond, candidate residues were needed to be replaced with Cysteine and as a result disulphide bond forms between sulphur atoms (Fig. 3.1).

In this project Disulfide by Design 2.11 web server is used to calculate potential residues (see appendix) which can be mutate to cysteine to form disulfide bond. After running the calculation, 5 mutations are selected based on preferred criteria (potential energy) (Table 3.1).

3.4 Molecular Dynamic Simulation Steps

Generally, there are three stages in molecular dynamics simulation: preparation of the input, production molecular dynamics, and analysis of the result (Fig. 3.2).

Fig. 3.1 Disulfide bond between to cysteine residues [1]

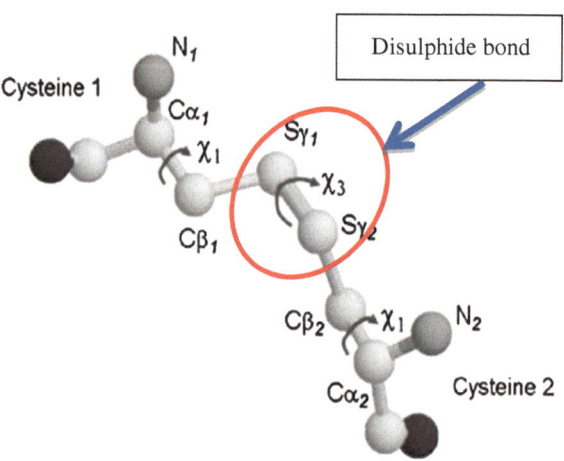

Table 3.1 The table illustrates files number corresponding introduced mutations

Mutant protein numbers	Mutations
1	E385C and A392C
2	Y321C and A333C
3	T383C and T399C
4	A187C
5	D257C and L346C

Fig. 3.2 Molecular dynamic simulation stages [2]

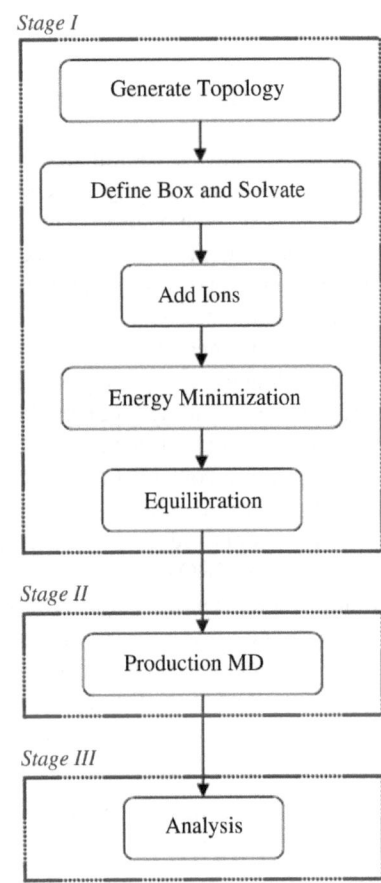

Stage I Preparation this stage has multiple steps including generating the topology file; defining a box and filling it with solvent, and adding any counter-ions to neutralize the system; performing energy minimization to provide stable simulation; performing equilibration for sufficient time to get stable pressure, temperature and energy.

Stage II *Production* this stage is the longest stage resulting in a trajectory containing coordinates and velocities of the system.

Stage III *Analysis* the last stage includes analysis of the resulting trajectory and data files to obtain information on the property of the molecule. Some important quantities calculated in this stage include RMS difference between two structures, RMS fluctuations, and rigidity or constant force, etc.

In this study whole simulation was carry out by using GROMACS machine version 4.5.5 (http://www.gromacs.org/).

3.4.1 pdb2gmx (Topology File Generation) Tool

In order to generate coordinates in GROMACS this tool reads a.pdb (or.gro) file, and then generate mainly GROMACS coordinate file (.gro) and topology file. These files can then be used to generate a run input file. The topology file (topol.top by default) contains all the data needed to explain the molecule in a simulation. This information includes force field specific non-bonded parameters (atom types, Van der Waals parameters, and charges) and bonded parameters (bond lengths, angles, dihedrals, and their force constants).

According to these principles, generated mutated pdb files are used to make three types of files including topology for the molecule and solvent, position restraint file and post-processed structure file (gro).

The pdb file contains one unnatural amino acids which the parameters was not configured in the GROMACS program, in order to tackle this obstacle the parameters obtained from webserver (http://viennaptm.univie.ac.at/) and implemented to the GROMACS database. The following command line is used to issue to this aim for pdb file number 1 and consequently for the rest of the files only the number in command line is changed. Once running this command some option is given for choosing force field so the modified GROMOS force field (GROMOS96 54a7 force field (Eur. Biophys. J. (2011), 40, 843–856, doi:10.1007/s00249-011-0700-9) contains all force-field parameters and building blocks to run simulations of post-translationally modified proteins is selected.

pdb2gmx -f 1.pdb -o 1_processed.gro –p 1.top

3.4.2 Unit Cell Defining the by the Tools, editconf and Adding Solvent

The editconf program is used to process a coordinate file further. In addition it's most common applications are defining a box size, specifying box type, locating the protein molecule in the centre of the box and most importantly scaling the box size to fix the density of the solvent. Therefore the command line below is used to define a cubic box.

editconf -f fws.gro -o box.gro -center x y z

As a result the program defines a cubic box and places the protein in the centre of the box. Then in order to add solvent 'genbox' program is used. Since in this case solvent is water the command line below is used which TI4P model of water is used.

genbox -cp fws.gro -cs spc216.gro -o fws_b4em.gro -p fws.top

3.4.3 Addition of Ions

In order to satisfy requirements of using long range electrostatic interaction algorithm (particale mesh Ewald summation), the net charge of the system needs to be set to zero.

In this case the System total charge was -23.220 so that 23 sodium ions were inserted into the simulation box. the program grompp first used to generate an atomic description of the system by using mdp file (molecular dynamic parameter file) and the command below is issued to assemble tpr file:

grompp -f em.mdp -c fws_b4em.gro -p fws.top -o ion.tpr -maxwarn 5
genion -s ion.tpr -o fws_b4em.gro -neutral -pname Na+ -np 23

As a result the net charge is not zero but it is acceptable which negative 0.220 is.

3.4.4 Energy Minimization as a Critical Step

In order to make sure that the system is relaxed and has no serious clashes or unsuitable geometry, the potential energy of the system needs to be minimized and the procedure called energy minimization (EM).

The procedure for EM is similar to the addition of ions. Grompp is used to assemble the coordinates, topology, and simulation parameters into a binary input file (.tpr), but this time, as a substitute of providing the.tpr to genion, the energy minimization is run over the GROMACS, using mdrun programme of the package. EM confirms the good quality initial structure for molecular dynamics simulation.

In this study maximum force was set in minim.mdp—"emtol = 1000.0"— indicating a target Fmax of no greater than $1{,}000$ kJ mol^{-1} nm^{-1}.

The binary input is gathered using grompp using following input parameter file:

grompp -f em.mdp -c fws_b4em.gro -p fws.top -o em.tpr -maxwarn 5

And then mdrun is evoked by command below:

mdrun -v -deffnm em

3.4.5 Equilibration of Temperature and Pressure

The equilibration was done by running the pre MD run by specific mdp file configuration. The following methods are used to equilibrate the system:

PME to treat coulomb potential. PME stands for particle mesh ewald electrostatics, which is the best method for computing long-range electrostatics (gives more reliable energy estimates especially for systems where counterions like Na+, The all-bonds option under constrains applies the Linear constraint algorithm for fixing all bond lengths in the system.

grompp -f pr.mdp -c em.gro -p fws.top -o pr.tpr -maxwarn 5
mdrun -v -deffnm pr

3.4.6 MD Production

Molecular dynamic simulation was performed using Gromacs 4.6.3 program [3] by modified GROMOS force field (GROMOS96 54a7 force field (Eur. Biophys. J. (2011), 40, 843–856, doi:10.1007/s00249-011-0700-9) implemented on a LINUX operating system. Prepared PDB file of six mutants and a native structure were used for MD simulation. After equilibration production MD was run for 20 ns at constant temperature and pressure. The LINCS algorithm [4] was used to constrain the bond length. Periodic boundary conditions were applied to the system. The electrostatics interactions were calculated using a PME algorithm with 0.9 Å cut off [5]. During the simulation every 1.000 ps of the actual frame was stored. The integration time step was 2 fs, with the neighbor list being updated every fifth step by using the grid option and a cut-off distance of 12 Å. The isotropic Parrinello-Rahman [6] protocol was used for pressure (1 bar), and the velocity-rescaling thermostat was used for temperature coupling. The trajectory data were calculate by GROMACS tools including g_rms, g_rmsd, g_gyrate and do_dssp visualized using XMGRACE program.

grompp -f md.mdp -c pr.gro -p fws.top -o md.tpr -maxwarn 3
mdrun -v -deffnm md

3.5 Tools for Analysis of Molecular Dynamics Trajectory

In this study, g_rmsd, g_rmsf and g_gyrate programs from gromacs utility are used in order to obtain respectively the root-mean-square deviation (RMSD), root-mean square fluctuation (RMSF), and radius of gyration.

The most frequently used measure for structure comparison in structural biology is, arguably, the atom-positional root mean-square deviation (RMSD) obtained after root-translational least-squares fitting. Its applications are diverse and include monitoring structural changes in simulations of protein folding and dynamics, assessing the quality of structure estimation schemes comparing the variety of model structures derived from experiments, assessing the properties of modeling approaches at unlike stages of resolution, and defining high-resolution shapes of polymers. Furthermore, structural diversity of an ensemble of bio molecular structures obtained through computer simulations is analyzed frequently by calculating an all-against-all distribution of RMSD values. In addition, in order to investigate the flexibility of protein structure RMSF is calculated. All runs have done and graphs are plotted by the GRACE plotting Linux software.

References

1. A.A. Dombkowski, Disulfide by DesignTM: a computational method for the rational design of disulfide bonds in proteins. Bioinformatics **19**(14), 1852–1853 (2003)
2. M. Yu, *Computational Modeling of Protein Dynamics with GROMACS and Java* (San Jose State University, San Jose, 2012)
3. S. Pronk, S. Páll, R. Schulz, P. Larsson, P. Bjelkmar, R. Apostolov, M.R. Shirts, J.C. Smith, P.M. Kasson, D. van der Spoel, B. Hess, E. Lindahl, GROMACS 4.5: a high-throughput and highly parallel open source molecular simulation toolkit. Bioinformatics **29**(7), 845–854 (2013)
4. B. Hess, H. Bekker, H.J.C. Berendsen, J.G.E.M. Fraaije, LINCS: a linear constraint solver for molecular simulations. J. Comput. Chem. **18**(12), 1463–1472 (1997)
5. T. Darden, D. York, L. Pedersen, Particle mesh Ewald: an N· log (N) method for Ewald sums in large systems. J. Chem. Phys. **98**(12), 10089 (1993)
6. M. Parrinello, Polymorphic transitions in single crystals: a new molecular dynamics method. J. Appl. Phys. **52**(12), 7182 (1981)

Chapter 4
Result and Discussion of Molecular Dynamic Simulation of Created Mutants

Abstract According to prediction nine residue 67, 187 ALA, 257 ASP, 346 LYS, 321 TYR, 333 ALA, 383 THR, 399 THR, 385 GLU and 392 ALA are selected to be mutated to cysteine, as a result 5 mutant are created. The result from molecular dynamic simulation trajectory calculations shows that mutations caused considerable effect on protein stability. RMSF calculation indicates that there are four flexible regions in the structure which mostly are loops. In mutant number 1 (E385C and A392C), mutation redacted flexibility of two regions and RMSD value shows lower value than the native one which means more stability where RG calculation suggest that structure tends to become more compact than native. In mutant number 2 (Y321C and A333C), the effect was more distinctive and the mutation caused more decrease in flexibility. In mutant number 3 (T383C and T399C), both two mutated residue located in loop region and mutation caused more stability in comparison to native structure like previous mutants. Although it caused reduction in flexibility of three region in protein stability it cause growth of flexibility in other region surprisingly. In mutant number 4 (A187C), mutation caused more stability according to RMSD RMSF and RG calculations. In addition mutation caused reduction in flexibility of four region notably. In mutant number 5 (D257C and L346C), where mutations introduce in loop regions, similar to other mutants graphs analysis show that stability of mutant is increased.

Keywords Mutation · Flexibility · Loop · Helix and compactness · Stability

4.1 Nominated Residues for Mutation

Based on the result coming from Disulfide by Design software, 5 pair residue are selected to be mutated to cysteine in order to crate disulfide bonds however one of the cysteine (Cys 67) was naturally existed and only the pair residue is mutated to cysteine. The 10 residue are highlighted by yellow in the Fig. 4.1.

© The Author(s) 2015
B. Barati and I. Sadegh Amiri, *In Silico Engineering of Disulphide Bonds to Produce Stable Cellulase*, SpringerBriefs in Applied Sciences and Technology, DOI 10.1007/978-981-287-432-0_4

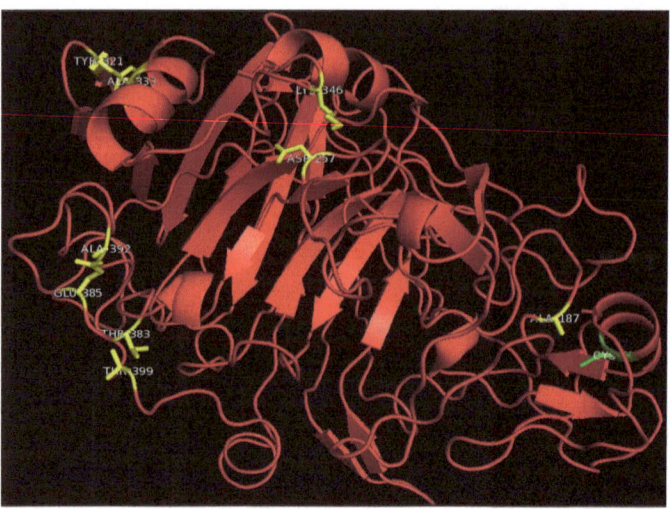

Fig. 4.1 All ten nominated residue 67 CYS, 187 ALA, 257 ASP, 346 LYS, 321 TYR, 333 ALA, 383 THR, 399 THR, 385 GLU and 392 ALA are highlighted by *yellow* and the native cysteine is highlighted by *green*

4.2 Native Protein Structure and Stability Analysis

The experimentally determined structure of native protein obtained from protein structure data bank (PDB) which naturally contains 10 native disulfide bridges. The obtained pdb file was loaded to the PyMol software (The PyMOL Molecular Graphics System, Version 1.3 Schrödinger, LLC.) to visualize the structure. The spheres in the Fig. 4.2 illustrate disulfide bonds.

In order to control the simulation for native protein with the same parameter and condition simulation was run. From resulted trajectory file RMSD, RMSF and Radius of gyration (Rg) have been computed.

During this period (Fig. 4.3). RMSD was started from 0.12 nm and ended to 0.325 nm. In this simulation RMSD calculation shows steady stability since there is no obvious fluctuation and the increasing is dramatically, it shows protein is in reasonable stability in the nature.

From Fig. 4.4 it can be seen that there are generally four sharp picks in the graph which illustrate meaningful flexibility of these region that may cause instability of the structure.

As shown in Fig. 4.5 highlighted region in the 3D structure are the flexible region based on the RMSF calculation, it is understood that these region are mostly loop and one of them is helix (Fig. 4.6).

Radius of gyration of the native protein is started from 2.053 and ended to 2.12 which means that around 0.1 nm the compactness is increased. This short change in RG means the protein remained relatively stable during the simulation since its

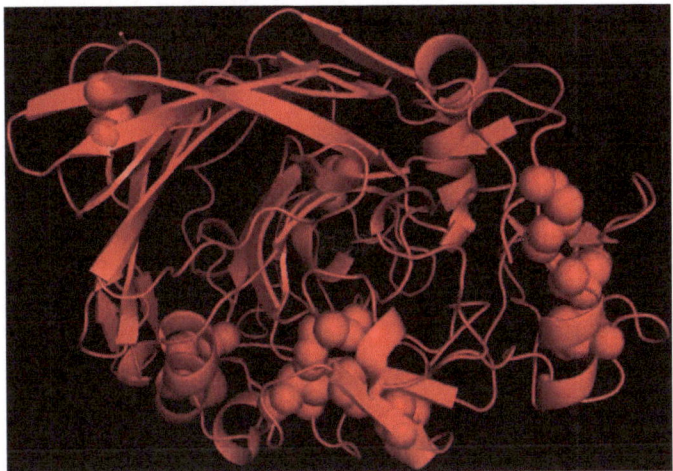

Fig. 4.2 The native protein structure with disulfide bonds. *Red* spheres are natural disulfide bonds

Fig. 4.3 RMSD calculation of native protein for the period of 20,000 ps (20 ns)

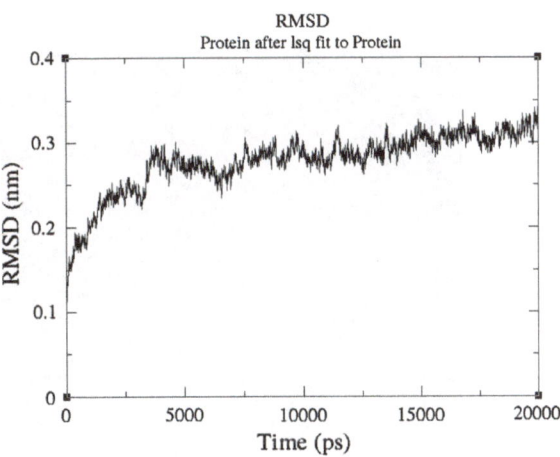

compactness is not changed meaningfully. Only around the time 3,500 ps one jump is occurred which caused the increase (Fig. 4.7).

4.3 Mutated Protein Number 1 (E385C and A392C) Structure and Stability Analysis

The residues Glu 385 and ALA 392 are mutated to cysteine and disulfide bond is formed as a result. The new disulfide bond is highlighted in the Fig. 4.8 by white colour. It is noticeable that these two residues were located in loop part of the protein and close to the surface.

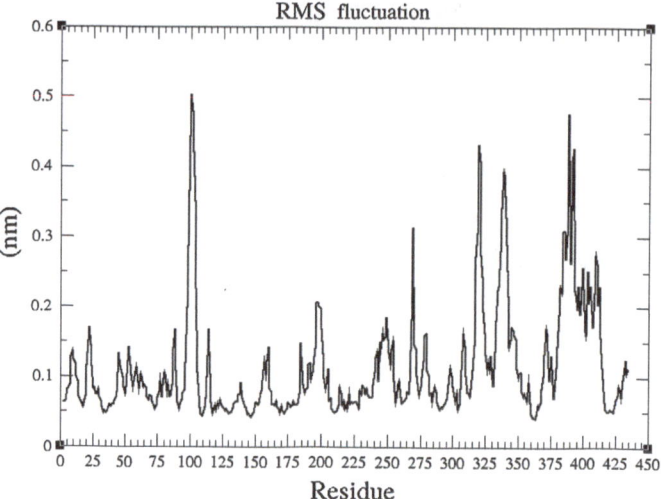

Fig. 4.4 RMS fluctuation of the native protein during simulation against carbon alpha

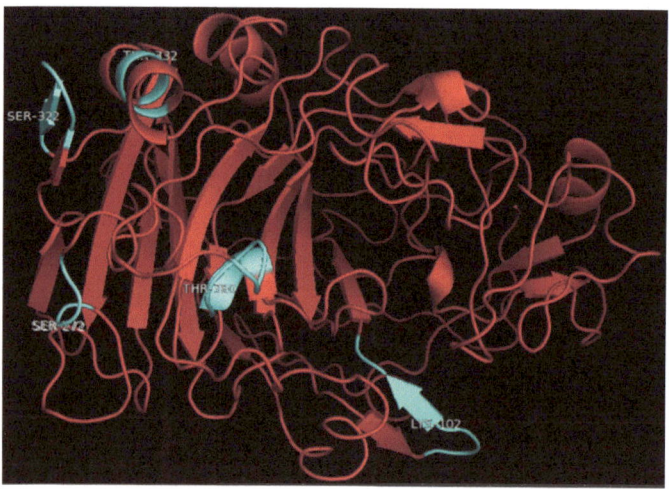

Fig. 4.5 Flexible region of protein based on RMSF calculation

A small jump can be seen in both result when for the native is earlier and occur near 3,000 ps and for the mutant occurs in 5,000 ps, then both protein spend the rest of the simulation in the steady mode. While the RMSD of native protein tends to increase the mutant remain in the same range. Although RMSD of both stricture start from the same value, the mutant structure tends to keep the RMSD value lower than native one.

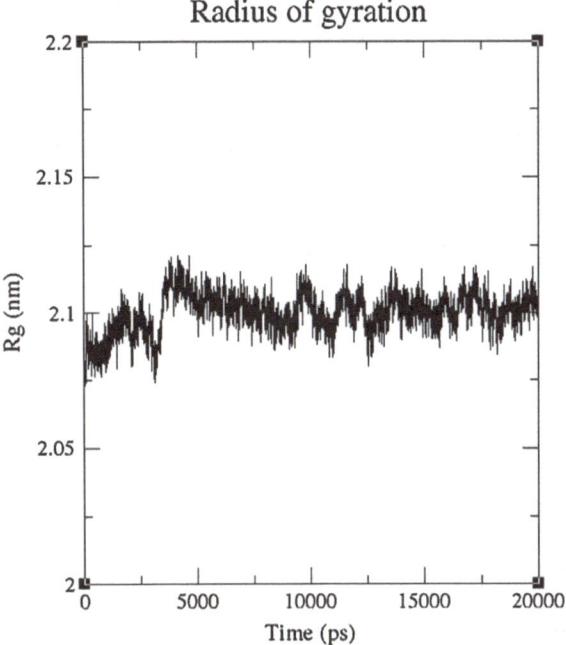

Fig. 4.6 Radius of gyration during 20,000 ps simulation

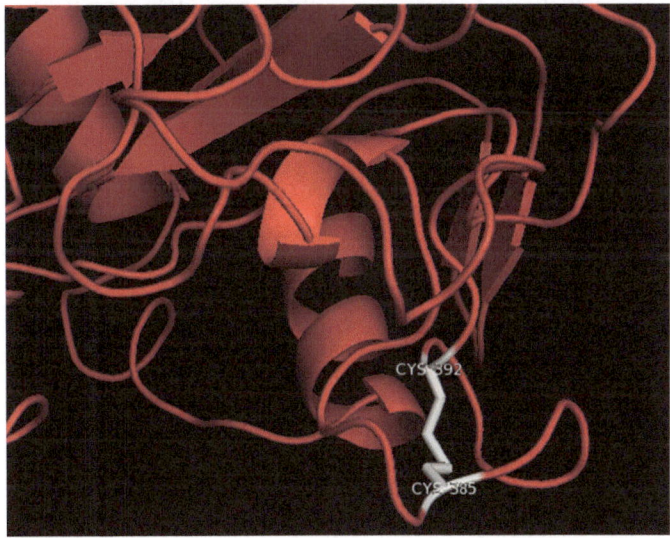

Fig. 4.7 Position mutated residues (Glu 385 and ALA 392) that formed new disulfide bridge in mutated protein

Fig. 4.8 RMSD calculations
for both native protein and
mutated protein number 1
(E385C and A392C) during
a period of 20,000 ps. The
black line shows native
protein RMSD and the *red
line* shows the RMSD of the
protein mutated

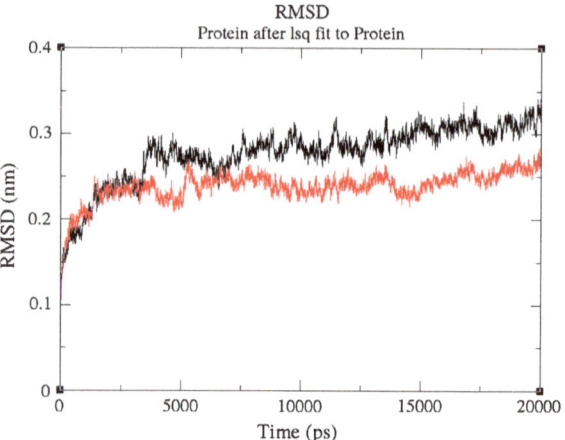

Fig. 4.8 RMSD calculations for both native protein and mutated protein number 1 (E385C and A392C) during a period of 20,000 ps. The *black line* shows native protein RMSD and the *red line* shows the RMSD of the protein mutated

Mutation location is highlighted by green circle in the graph which it can be seen that mutation decreased the flexibility of the location as well as the other part which is highlighted by red circle which flexibility greatly is decreased. However there are two location which is pointed by blue arrows that flexibility slightly increased (Fig. 4.9).

From the comparative graph of the Fig. 4.10 it can be seen that radius of gyration for the mutated protein (E385C and A392C) is decreased obviously during simulation. Although both started at the same value, they begin the separate from the time 4,000 ps when the native structure begin to expend for a while then keep the

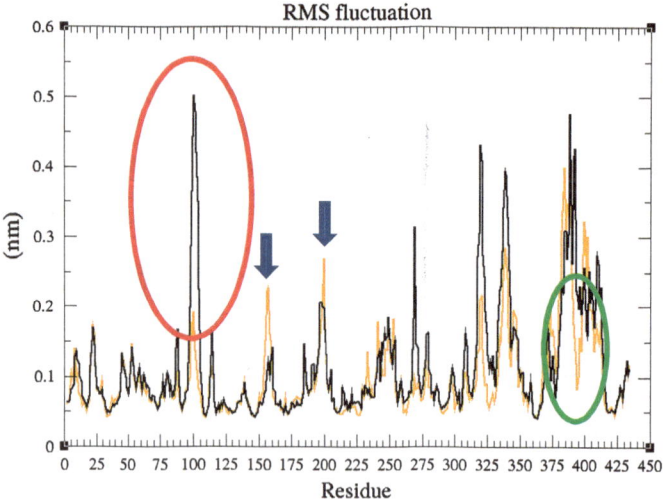

Fig. 4.9 RMSD fluctuations of the mutated protein number 1 (E385C and A392C) and native. The *yellow line* and *black line* shows RMSF for mutated and native protein respectively

Fig. 4.10 Radius of gyration of the mutated protein 1 (E385C and A392C) and native protein. The *black* and *red lines* indicate values for native and mutated protein respectively

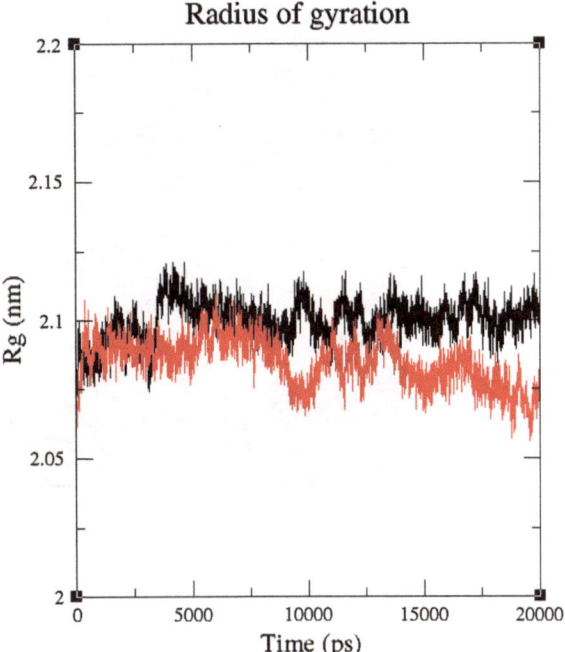

expanded structure in the same value to the end. While the mutated protein began to compact .at the end of the simulation time a meaningful gap can be seen which indicates the mutated structure is more compacted and stable than the native one.

4.4 Mutated Protein Number 2 (Y321C and A333C) Structure and Stability Analysis

The second mutated protein is produced by mutating TYR 321 and ALA 333 to cysteine, TYR 321 is located on the small beta sheet and ALA 333 is located on the alpha helix. The new disulfide bridge is formed and highlighted by magenta colour and is shown in the Fig. 4.11.

From the Fig. 4.12 it can be seen that both structure spend the same situation for the first 3,000 ps then RMSD of native structure jump from 0.25 to 0.3 nm and tends to increase slightly while for the mutated protein fluctuations are less and after 10,000 ps the structure become stable at the RMSD value of 0.25 nm.

From calculated RMSF graph it can be seen from Fig. 4.13 that as a result of introduced mutation, flexibility of three region slightly decreased and pointed by yellow and green circles which the green circle is the location of mutations. It can be seen that in the introduced mutation location, the flexibility successfully redacted. The overall flexibility of the rest of the residue is decreased or remained at

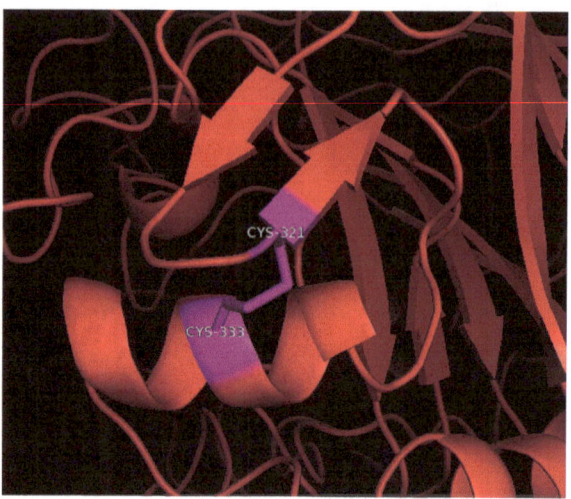

Fig. 4.11 TYR 321 and ALA 333 are mutated to cysteine

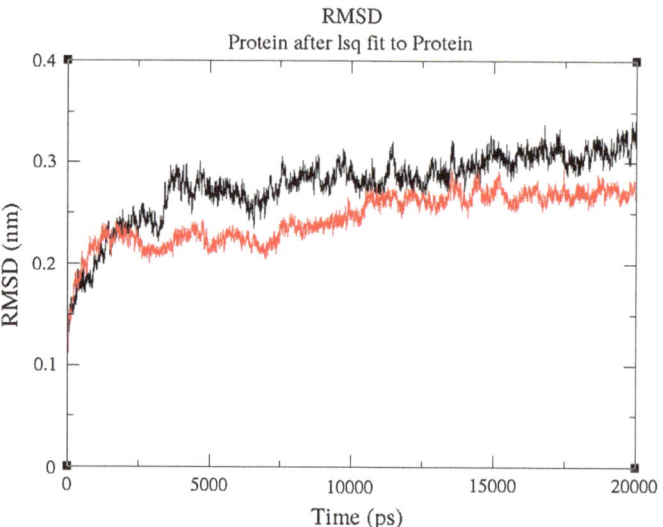

Fig. 4.12 RMSD calculations for both normal and mutated protein (Y321C and A333C), the *black line* and *red line* show RMSD values for native and mutated protein respectively

the same degree. All together the residue flexibility decreased and mutation is not caused more flexibility in other parts.

It can be seen from Fig. 4.14 that there is obvious decrease in radius of gyration of mutated protein compared to the native structure, the mutated protein during simulation time start to become more compact.

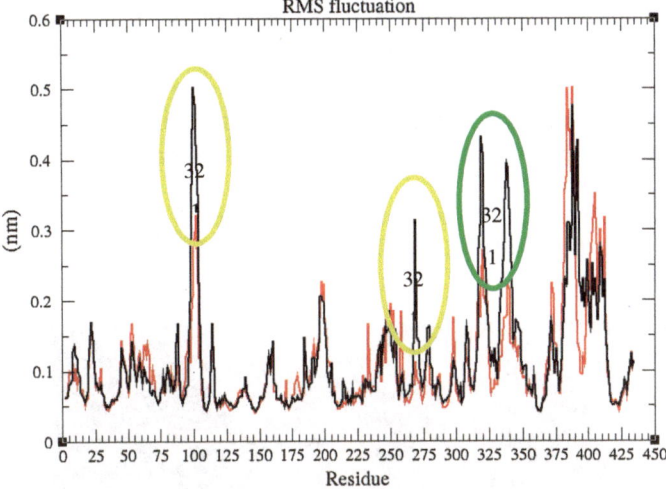

Fig. 4.13 RMSF of the mutated protein 2 (Y321C and A333C) which is highlighted by *red color* and RMSF of the native protein by *black color* for the 446 residues during 20,000 ps simulation

Fig. 4.14 Radius of gyration for mutated protein (TYR 321 and ALA 333) and the native protein. *Red line* and *black line* refer to mutated protein and native protein respectively

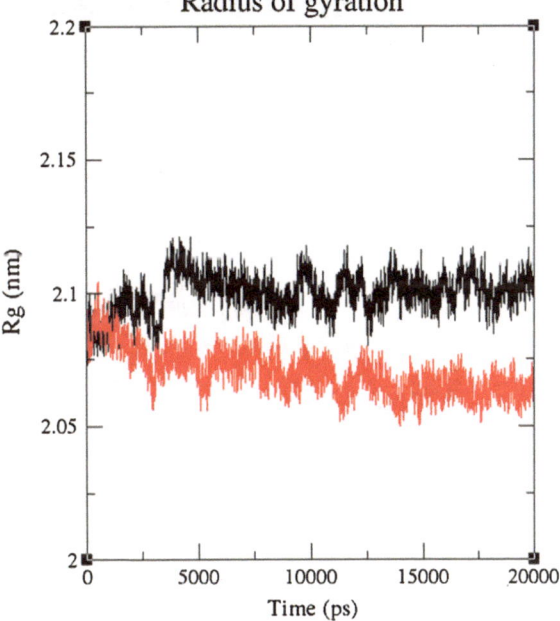

4.5 Mutated Protein Number 3 (T383C and T399C) Structure and Stability Analysis

The third protein is resulted from mutation of residue 383 THR, 399 THR which both are located the loop of the protein. The resulted disulfide bridge is shown by cyan colour in the Fig. 4.15.

From Fig. 4.16 it can be seen that RMSD values of mutated protein from the starting point of the simulation remained the same values with native protein until

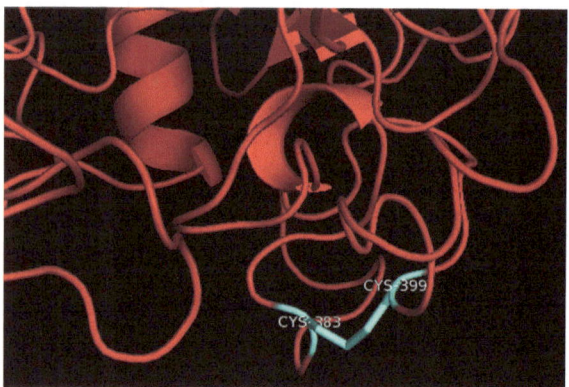

Fig. 4.15 Highlighted *cyan* residues in the figure show the residues that are mutated

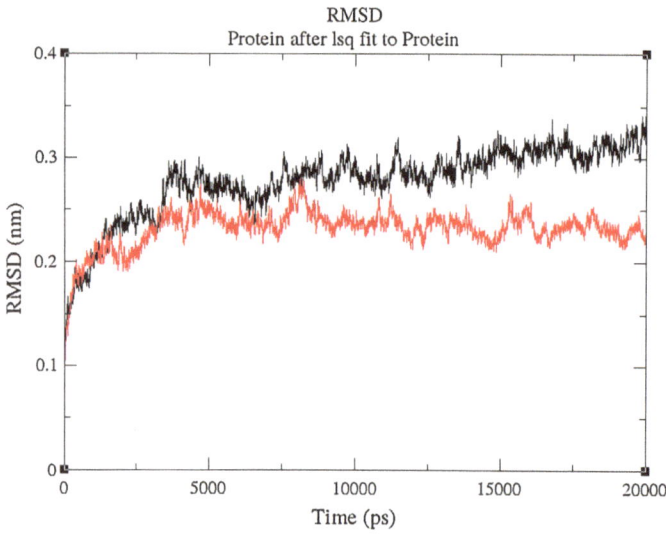

Fig. 4.16 RMSD calculations for both native protein and mutated protein (T383C and T399C). The *black line* and *red line* show RMSD values for native and mutated protein respectively

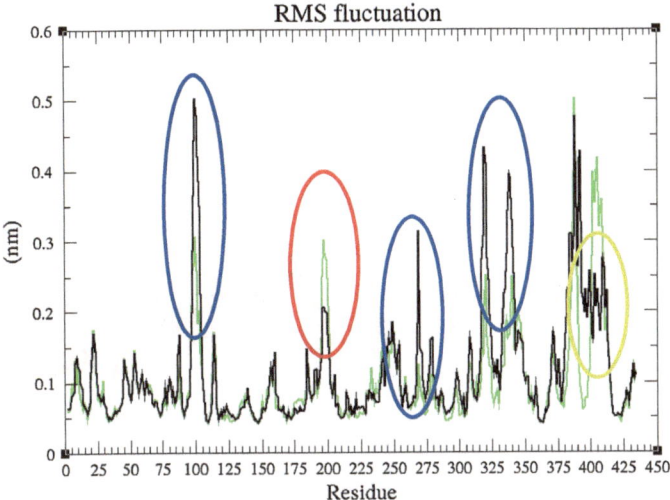

Fig. 4.17 RMSF of normal and mutated protein 3 (T383C and T399C). The *black line* and *green line* show the values for native and mutated protein respectively

2,000 ps then a sharp increase is happened in RMSD value of the native following to that it endure steady increasing and reached to more than 0.3 nm while the mutated protein after 5,000 ps remain steady until the end of the simulation and RMSD reached to slightly more than 0.2 nm.

From the RMSF graph Fig. 4.17 it can be seen that in five region the flexibility notably decreased (blue circles) and yellow circle (mutation sites) while only in the one region the flexibility increased pointed by red circle. In this mutant more reduction in flexibility can be seen compared to the previous mutant.

From Fig. 4.18 it can be seen that the radius of gyration values for both proteins remain roughly the same at the beginning and towards the middle of the simulation. As a result of mutation the compactness start increasing from 12,000 ps to the end of the period.

4.6 Mutated Protein Number 4 (A187C) Structure and Stability Analysis

The third protein is resulted from mutation of residue 67 CYS, 187 ALA which CYS 67 is located I alpha helix and the residue ALA 187 in loop. The resulted disulfide bridge is shown by blue colour in the Fig. 4.19.

Form the Fig. 4.20 it can be seen that mutant and native for the first 2,000 ps follow similar pattern of RMSD fluctuation then RMSD of native jump to near 0.3 and for the rest of the simulation increasing dramatically and reach to 0.325. While

Fig. 4.18 Radius of gyration
of the mutated protein 3
(T383C and T399C) and
normal the *black* and *red lines*
refer to normal and mutated
protein respectively

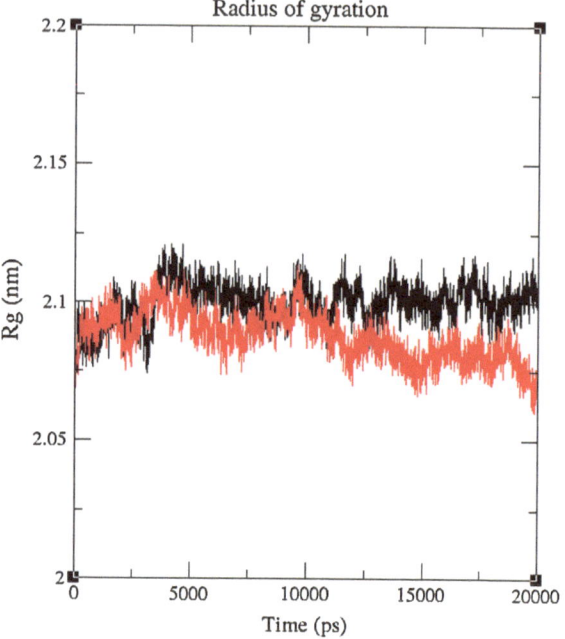

Fig. 4.18 Radius of gyration
of the mutated protein 3
(T383C and T399C) and
normal the *black* and *red lines*
refer to normal and mutated
protein respectively

Fig. 4.19 Mutated protein,
the highlighted *blue* residue
are new residue result of
mutation in 67 CYS, 187
ALA

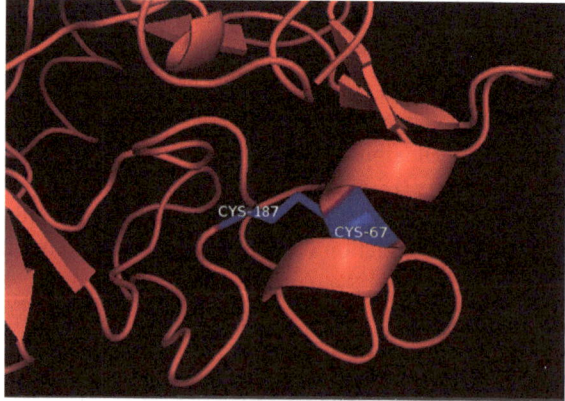

for the mutant instate of jump there is increasing by moderate slope which increase
the value from 0.2 to 0.25 and for the rest of the simulation the RMSD fluctuate at
the same level.

From the RMSF graph Fig. 4.21 although it can be seen that in five region the
flexibility notably decreased (green and red circles), the flexibility of mutation site
redacted but not significantly (red circle). In this mutant more reduction in flexibility can be seen compared to the previous mutant. It seems that mutation effected
on the other part more significantly than the other parts.

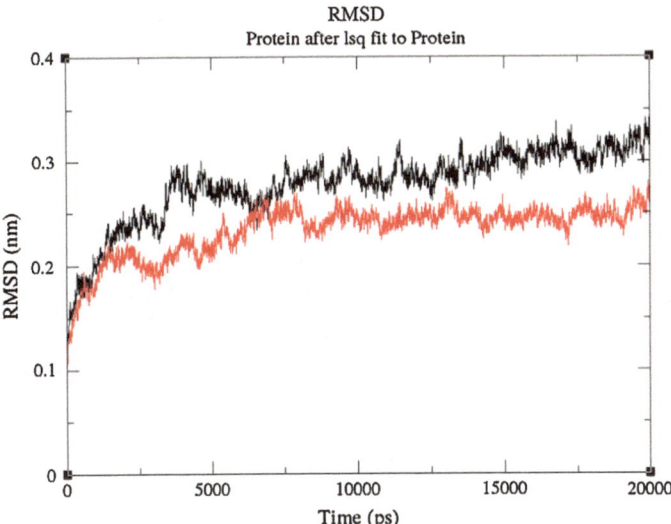

Fig. 4.20 RMSD calculations for mutated protein 4 (A187C) and native protein. The *black* and *red line* refer to native protein and mutated protein respectively

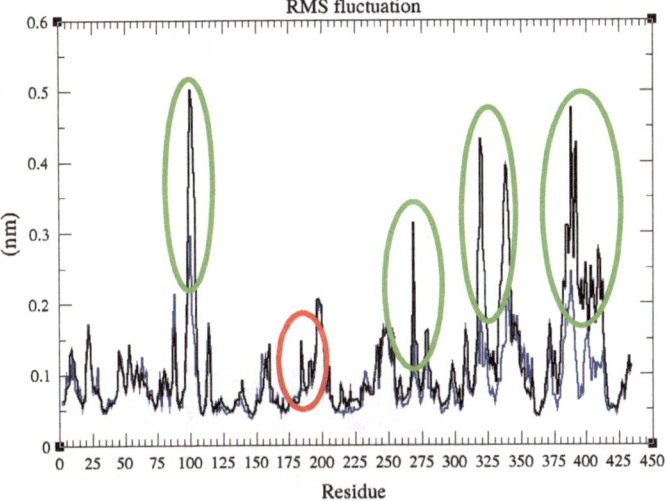

Fig. 4.21 RMSF for mutated protein 4 (A187C) and native protein. The *black* and *blue lines* show RMSF of the native and mutated protein respectively

From the RG calculation it can be seen from the Fig. 4.22 that towards the end of the simulation the mutated protein becomes more compact even more than native protein which means more stability in mutated protein. Although both structure show the same pattern of fluctuation in the graph.

Fig. 4.22 Radius of gyration of the mutated protein 4 (A187C) and the native protein the *red color* and *black color* refers to the RG values of the mutated protein and native protein respectively

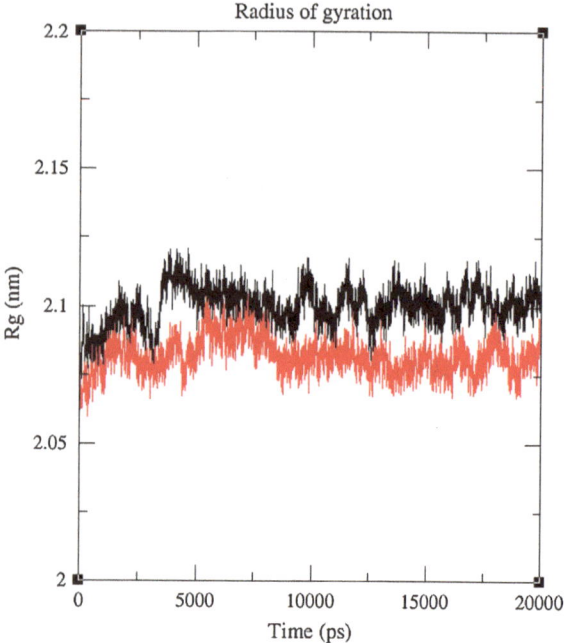

4.7 Mutated Protein Number 5 (D257C and L346C) Structure and Stability Analysis

In order to create mutant number 5 the residue 257 ASP and 346 LYS are mutated to cysteine and formed disulfide bond. Both residue were located in the loop of the protein (Fig. 4.23).

It can be seen from the Fig. 4.24 that from around time 1,000 ps the graph begin to separate the mutant tend to become more stable compared to native one, the

Fig. 4.23 The highlighted disulfide bond by *light yellow* is formed by mutation of 257 ASP and 346 LYS

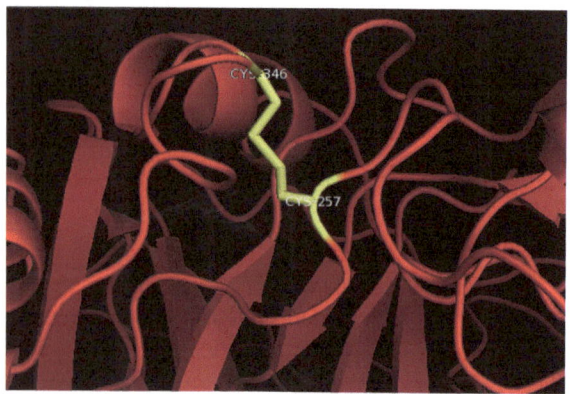

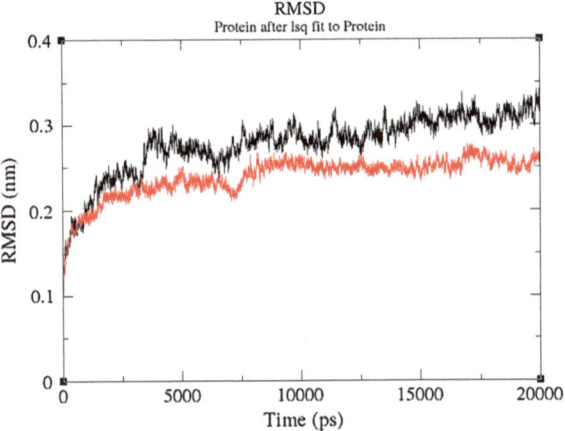

Fig. 4.24 RMSD mutated protein 5 (D257C and L346C) by *black line*

RMSD value of mutant remain around 0.25 while the native stricture tend to be slightly instable by increasing RMSD value which reached to the maximum of 0.31.

Red circled regions (Fig. 4.25) are the location that the most changes in flexibility occurs these changes are the result of the introduced mutations. As can be seen in the pointed region the flexibility of the mutant is decreased and resulted more rigid structure. The mutation sites are pointed by blue arrow, although mutations effected in some other parts notably the mutation site did not effected much.

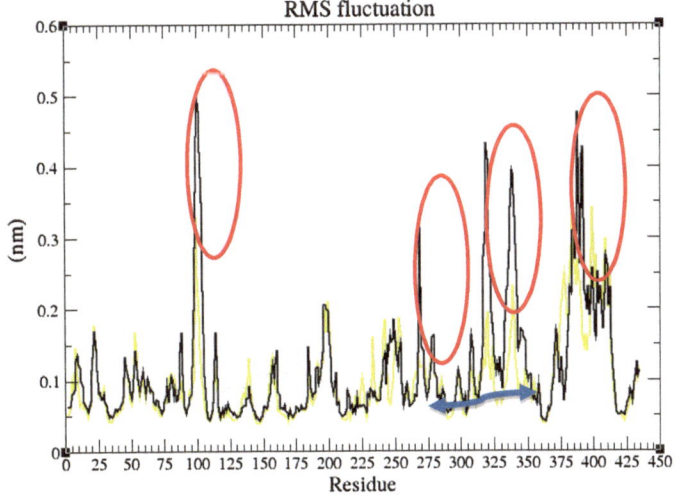

Fig. 4.25 RMSF of all mutated proteins 5 (D257C and L346C) and native protein *yellow line* mutant and the *black line* native

Fig. 4.26 Radius of gyration
of the all mutated protein 5
(D257C and L346C) and
native protein

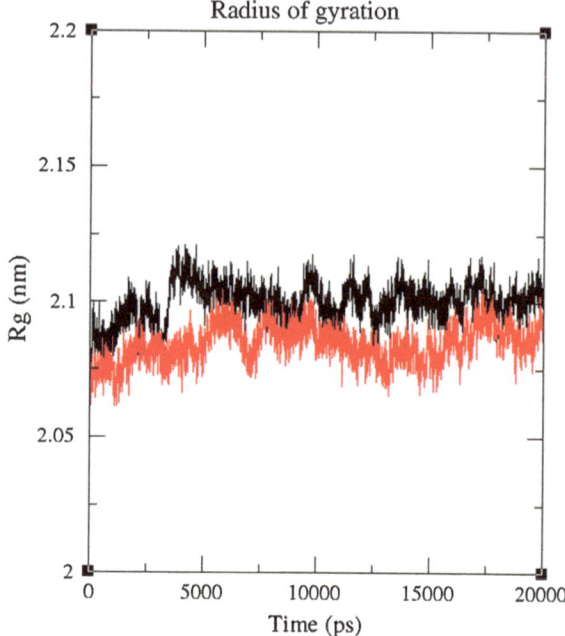

Fig. 4.26 Radius of gyration of the all mutated protein 5 (D257C and L346C) and native protein

From Fig. 4.26 it can be seen that, mutated protein 5 (D257C and L346C) and native proteins are at roughly the same level of the compactness since both end up to the approximately at same value which is near 2.1 nm. Although during the period of mutation the mutant structure was more compact.

Chapter 5
Conclusion of Simulation Analysis of Mutants

Abstract Introduced disulfide bonds by mutating nine residue to cysteine all caused more stability to protein structure as all RMSD values were less than native protein. In addition mutations caused more rigidity in the structure since the RMSF values reduced in several regions in all the mutants. Moreover the compactness of protein increased as radius of gyration in all the mutant was less than native structure. Although this mutant suggesting more stability in structure, still other research can be done such as combination of this mutant and seeing the accumulation effect of disulfide bonds, running simulation in longer time to more accurate analyses, the algorithm to predict candidate residue for mutation can be improved.

Keywords Compactness · Rigidity computational algorithm · Combination of mutant · Accumulation effect

5.1 Conclusion of MD Simulation of Created Mutants

Protein structure can be consecrated as an architectural flexible structure, in the same way that welding can improve the stability of the structures, disulfide bonds can as well. Based on RMSD graph analysis, all the mutation were stabilizing which lowered RMSD and kept the value in same range until end of simulation time. Although disulfide bonds are stabilizing bonds by decreasing the flexibility of the structure, the effect is not always reduction in flexibility, to be more exact when disulfide bridges are added to the structure the flexibility of all residue may or may not be reduced depending on the mutations sites, as in some rare cases mutations caused more flexibility of some residue. According to RG analysis, in all the mutant compactness reduced which indicates the structure become tighter, as a matter of fact the structure that is more stable and compacter is more stable in higher temperature rather than native protein which confirms RMSD calculations. Therefore it can be concluded that these computationally mutated proteins are more thermostable than native protein. By analyzing result of the all runs of the simulation it can

© The Author(s) 2015

B. Barati and I. Sadegh Amiri, *In Silico Engineering of Disulphide Bonds to Produce Stable Cellulase*, SpringerBriefs in Applied Sciences and Technology, DOI 10.1007/978-981-287-432-0_5

be concluded that introducing Disulfide Bridge predicted by **Disulfide by Design** software were computationally successful in order to improve stability of the enzyme.

5.2 Future Work of the Disulfide Bond Engineering to Cellulase

The following future works, may be proposed:

1. Second developing algorithms to improve accuracy for prediction of the potential residue that can be mutated to form disulfide bridges.
2. First running simulation for longer time to make sure the conformation are in equilibrium conformation.
3. Third introducing more disulfide bridges and doing more simulation to find better (more stable) mutated protein and creating combination of mutations to see the effect of combinations.
4. Performing protein engineering in wet lab by introducing the most stabilized mutations in order to see the result in the reality.

Acknowledgements Iraj Sadegh Amiri would like to acknowledge the financial support from University Malaya/MOHE under grant number UM.C/625/1/HIR/MOHE/SCI/29 and RU002/2013.